Shaping China's Innovation Future

ELGAR INTELLECTUAL PROPERTY AND GLOBAL DEVELOPMENT

Series Editor: Peter K. Yu, *Kern Family Chair in Intellectual Property Law and Director, Intellectual Property Law Center, Drake University, USA*

Rapid global economic integration and the increasing importance of technology and information goods have created the need for a broader, deeper and more critical understanding of intellectual property laws and policies. This uniquely-designed book series provides an interdisciplinary forum for advancing the debate on the global intellectual property system and related issues that intersect with transnational politics, international governance, and global economic, social, cultural and technological development. The series features the works of established experts and emerging voices in the academy as well as those practising on the frontlines. The series' high-quality, informed and accessible volumes include a wide range of materials such as historical narratives, theoretical explanations, substantive discussions, critical evaluations, empirical analyses, comparative studies, and formulations of practical solutions and best practices. The series will appeal to academics, policy makers, judges, practitioners, transnational lawyers and civil society groups as well as students of law, politics, culture, political economy, international relations and development studies.

Titles in the series include:

Intellectual Property and Sustainable Development
Development Agendas in a Changing World
Edited by Ricardo Meléndez-Ortiz and Pedro Roffe

Internet Domain Names, Trademarks and Free Speech
Jacqueline Lipton

Shaping China's Innovation Future
University Technology Transfer in Transition
John L. Orcutt and Hong Shen

Shaping China's Innovation Future

University Technology Transfer in Transition

John L. Orcutt

Associate Dean for Faculty Research and Professor of Law, University of New Hampshire School of Law – Franklin Pierce Center for Intellectual Property, USA

and

Hong Shen

Partner, Longan Law Firm, China

ELGAR INTELLECTUAL PROPERTY AND GLOBAL DEVELOPMENT

Edward Elgar

Cheltenham, UK • Northampton, MA, USA

Published by
Edward Elgar Publishing Limited
The Lypiatts
15 Lansdown Road
Cheltenham
Glos GL50 2JA
UK

Edward Elgar Publishing, Inc.
William Pratt House
9 Dewey Court
Northampton
Massachusetts 01060
USA

A catalogue record for this book
is available from the British Library

Library of Congress Control Number: 2010927665

MIX
Paper from
responsible sources
FSC® C018575

ISBN 978 1 84980 358 8 (cased)

Typeset by Servis Filmsetting Ltd, Stockport, Cheshire
Printed and bound by MPG Books Group, UK

Contents

Preface   Will China be the next technology superpower?

'Science is the first productive force.' – Deng Xiaoping's motto

Whenever we explain to anyone that we are working on a project involving China's innovation capacity, the first question we get back is invariably: will China be the next great science and technology (S&T) superpower? Almost unthinkable a mere 20 years ago (and maybe even 10 years ago), such a proposition is a very real possibility for the China of today. Since the late 1970s, China has gone through an enormous transformative period. The transformation that tends to draw the most attention has been China's stunning economic growth. Over the last three decades, China's economy has grown at an average annual rate of close to 10% – which is one of the highest periods of sustained economic growth in human history. During this relatively short period of time, China has grown from one of the poorest countries in the world in the 1970s into the world's third largest economy in 2009 – trailing only the United States and Japan.[1]

During this transcendent growth period, China has also experienced a more subtle, but equally important transformation. While the world continues to focus on China's economic growth and the fact that China has become the world's preferred destination for labor-intensive, low technology manufacturing, China's leadership has been laying the groundwork for the country to become a more innovative nation that can grow its economy as much by the strength of its S&T capabilities as it has so far done by its pool of cheap and productive labor. Over the last decade in particular, China has succeeded in dramatically improving the country's S&T capabilities. Research and development (R&D) expenditures have increased by several orders of magnitude, and similar increases can be found in China's S&T workforce, its export of high-tech products, its patent activity, and in the scholarly output of its scientific researchers (see Table P.1, which provides select S&T indicators for China from 1998 through 2007).

In addition to these impressive statistical data, anecdotal examples of China's growing S&T prowess abound in the areas of space technology, biotechnology (including notable achievements in genomics and stem cell

Table P.1   Select S&T indicators for China[1]

	1998	2001	2004	2007
1. R&D Expenditures				
Gross domestic expenditures on R&D (GERD) (RMB in billions)	55.1	104.2	196.6	371.0
R&D intensity (GERD/GDP)	0.70%[2]	0.95%[3]	1.23%	1.49%
Government S&T appropriation (GSTA) (RMB in billions)	43.9	70.3	109.5	211.4
GSTA as % of total government spending	4.1%[4]	3.7%[5]	3.84%	4.25%
GSTA/GERD	79.7%	67.5%	55.7%	57.0%
2. Human Resources in S&T				
(tens-of-thousands of full-time equivalents)				
Scientists and engineers	n/a[6]	74.3	92.6	142.3
R&D personnel	n/a	95.7	115.3	173.6
S&T personnel	n/a	314.1	348.1	454.4
3. High Technology				
Exports of high-tech products (450 in billions)	20.3	46.5	165.4	347.8
Share in total exports	11.0%[7]	17.5%[8]	27.9%	28.6%
4. Other Output Indicators (in thousands)				
Patent applications filed in China	122	204	354	694
Patents granted in China	68	114	190	352
S&T papers indexed by SCI, EI and ISTP[9]	35[10]	65[11]	111	208

Sources:
1. Ministry of Science and Technology of the People's Republic of China (MOST) (2008), *China S&T Statistics Databooks* unless otherwise indicated by a specific footnote. MOST (2008), *China S&T Statistics Databook* is available at http://www.sts. org.cn/sjlel.kjtjdt/data2008/cstsm08.htm. MOST (1998–2007), *China S&T Statistics Databooks* are available at www.most.gov.cn/eng/statistics/2007/index.htm.
2. MOST (2004), *China S&T Statistics Databook.*
3. MOST (2007), *China S&T Statistics Databook.*
4. MOST (2004), *China S&T Statistics Databook.*
5. MOST (2007), *China S&T Statistics Databook.*
6. The categories the Chinese government used for tracking human resources in S&T changed in 2006. The 1998 figures have not been converted into the new categories.
7. MOST (2004), *China S&T Statistics Databook.*
8. MOST (2007), *China S&T Statistics Databook.*
9. SCI stands for Science Citation Index, EI stands for Engineering Index, and ISTP stands for Index to Scientific & Technical Proceedings.
10. MOST (2004), *China S&T Statistics Databook.*
11. MOST (2007), *China S&T Statistics Databook.*

research), information technology, nanotechnology, and advanced energy technology.

- China is one of only three countries to put a person in space with its own rockets (and China recently conducted its first spacewalk).
- Chinese research teams helped to map the genome for rice and have since helped to extend genomic sequencing to other plants, as well as a variety of insects and parasites.
- China passed the United States as the leading exporter of information-technology goods in 2004.
- China has become a world leader in the field of nanotechnology – producing major nanotechnology breakthroughs (e.g., improved production of carbon nano-tubes) and generating a significant portion of the world's nanotechnology publications and patents and new nanotechnology firms.
- China has long been a leader in nuclear technology and is positioned to become a leader in a number of other energy fields, including clean coal and hydropower.

China's aspirations for the future are even greater than its past accomplishments. In 2006, China's leadership set formal goals for China to become an 'innovative nation' by 2020, which includes dedicating 2.5% of the country's GDP to R&D, and a world leader in S&T by the middle of the 21st century.[2]

How strong are China's technology capabilities? Can China really become an S&T superpower? These questions tend to spark a fair amount of debate, with the two sides frequently reaching almost polar opposite conclusions about China's S&T strength and future. One commentator likened the debate to two people holding the same coin, but with each person only looking at one side of the coin.[3] On the one hand, there is a pile of data that shows China has increased its output in several key S&T indicators. For example, China has dramatically increased the number of patents, scientific articles and engineers that it produces. China has also progressed in developing a high-tech manufacturing sector. This progress has led some to believe that China will soon overwhelm the rest of the world in technology. On the other hand, there is an equally large pile of data that suggests China's current technology capabilities are not that strong, and may remain weak for the foreseeable future. Much of China's progress in patents, scientific articles and engineer formation could be described as involving improvements in 'quantity,' but not necessarily 'quality.' In addition, China's improvements in higher-technology manufacturing remain overly dependent on foreign technology transfer, as

China has yet to develop domestic technology generation capabilities that truly rival those of the most developed countries.

These conflicting viewpoints beg the question – which side of the coin provides the more accurate description of China? Such a question does not lend itself to a simple black-or-white answer. It appears that there is truth on both sides of the debate. China's innovation system possesses a number of powerful strengths, but it also suffers from significant problems. Such a combination, not surprisingly, tends to generate rather uneven S&T results that provide ammunition to both sides of the debate.

While we are not in a position to predict whether China will be the next S&T superpower, we do feel quite confident in making a related projection. If China is to have a reasonable shot at becoming an S&T superpower, its university sector will need to be a major driver of the effort and will need to forge much stronger and more productive partnerships with its business sector. This book explores the rapidly evolving role of universities in China's innovation system and considers what China can do to ensure that its universities are well positioned to create valuable, commercial technology that can serve as the bedrock for China's future economic growth.

Acknowledgements

We could not have completed this project without our colleagues at the International Technology Transfer Institute (ITTI), which is part of the Franklin Pierce Center for Intellectual Property at the University of New Hampshire School of Law. ITTI is working to develop expertise and services in international technology transfer that could hold the key for the developing world's future economic development. Our ITTI colleagues, in particular Dr. Stanley Kowalski, Professor Karen Hersey and Asst. Professor Jon Cavicchi, who is the best intellectual property librarian in the world, have generously given us their time to discuss the various ideas in the book, review drafts and give us their valuable insights on the broad range of topics this book addresses.

Other colleagues that have been with us from the outset, and supported us through the trials and tribulations of writing this book by providing their comments, insight and general good cheer, include Professors William Murphy, Susan Richey, Mary Wong and Bill Hennessey of the University of New Hampshire School of Law – Franklin Pierce Center for Intellectual Property. Asst. Professor Barry Shanks, also from UNH Law, provided us with his always valuable library research support. They are all friends in the fullest sense of the word.

This book draws heavily from Hong's 30+ years of experience in the Chinese legal system. Her interactions with classmates and professors from the Peking University Law School in the late 1970s and early 1980s, the students she taught for more than a decade as a law professor at the China University of Political Science and the Law, the Chinese judges she clerked for, the Chinese attorneys she has worked with, and the clients she has served have all shaped our research and analysis in this book. Hong would particularly like to thank Ms. Huang Ermei, the Grand Justice of the Supreme People's Court of the People's Republic of China, and Mr. Jason Lee, her partner with the Long An Law Firm and a former Justice of the Supreme People's Court, who shared with her their valuable experience on the changes that have been taking place in China's court system and the growing prominence of the law in China.

We also want to thank the very talented Chinese university technology transfer experts who spent time talking with us about their experiences in making technology commercialization a more meaningful part of Chinese economic development.

During our work on this book, we have had a number of talented research assistants who have provided us with thoughtful and generous assistance in writing and editing this book. They include Dr. Kimberly Peaslee, Nathan Greene, John Kenyon and Nathalia Pence.

Finally, and most importantly, our families provided immeasurable support while we wrote (and rewrote) the manuscript. We appreciate their patience and understanding as we disappeared for large periods of time to work on the book. John would like to thank his beautiful and almost perfect wife, Corinne, and his three children, Xavier, Alexandre and Morgane, for everything they give to him. John particularly wants to thank Morgane for always remembering to ask him how the book was going and telling him things would be okay after bad days – it meant a lot. Hong would like to thank her wonderful parents for all the love they have given to her throughout her life. Her great father Shen Dingyi and her late mother Xie Linfeng have always been her strongest supporters. Hong thinks her mother would have enjoyed reading this book.

Abbreviations and acronyms

AIC	Administration for Industry and Commerce
AUTM	Association of University Technology Managers
CAS	Chinese Academy of Sciences
CKZTC	China Ke Zhao Tech Co., Ltd.
CNTVCC	China New Technology Venture Capital Company Ltd.
CVC	Corporate-backed venture capital firm
DEN	Dartmouth Entrepreneurial Network
ERISA	Employee Retirement Income Security Act of 1974 (U.S. law)
FDI	Foreign direct investment
FTP	Foreign technology purchases
FVC	Foreign venture capital firm
GANB	Generally accepted norms of behavior
GEM	Growth enterprise market
GERD	Gross domestic expenditures on R&D
GPCL	General Principles of Civil Law
GRI	Government research institute
GSTA	Government S&T appropriations
GVC	Government-backed venture capital firm
IP	Intellectual property
IPO	Initial public offering
JSLC	Joint stock limited company
Local IPOs	Local intellectual property offices
NCAC	National Copyright Administration of China
NPC	National People's Congress
NSTP	National Science and Technology Program
NTE	New technology enterprise
OECD	Organization for Economic Co-operation and Developmant
PRB	Patent Reexamination Board of SIPO
R&D	Research & development
RMB	Renminbi
SAFE	State Administration of Foreign Exchange
SDPC	State Development Planning Commission
SIPO	State Intellectual Property Office

SOE	State-owned enterprise
SSTC	State Science and Technology Commission
S&T	Science and technology
TTO	Technology transfer office
TVEs	Town and Village Enterprises
TRIPS	Agreement on Trade Related Aspects of Intellectual Property Rights
USU	University-owned start-up
UVC	University-backed venture capital firm
WIPO	World Intellectual Property Organization
WTO	World Trade Organization

1. Universities, technology commercialization and innovation systems

[T]he balance between knowledge and resources has shifted so far toward the former that knowledge has become perhaps the most important factor determining the standard of living – more than land, than tools, than labor.
– Paul S. Romer[1]

In today's knowledge-based world, the creation and commercialization of new technologies are the driving force behind sustainable economic growth and social prosperity. New technologies create new products, new markets, new processes for doing business, and even new industries, while improving the economy's overall efficiency and competitiveness. Reduced to its simplest concept, economic activity involves taking physical resources and rearranging them into things (products or services) that are more valuable than before. Because physical resources are scarce, economic growth requires more than just rearranging more physical resources, but eventually requires rearranging the physical resources better.[2] That is where innovation and technological advances come into play. Because knowledge can be shared and infinitely reused at little to no cost, technological advances are generally much more valuable than the underlying physical resources that they seek to transform.[3] As a result, those countries with the strongest innovation capabilities also tend to have the strongest and most advanced economies.

Universities are one of the most important elements in shaping a country's innovation capabilities. At the most basic level, universities help to develop the most technologically advanced portion of a country's workforce. A country cannot develop a consistently successful, technology-based economy without a critical mass of highly-skilled workers that are trained to operate in science and technology fields. Universities help to educate the scientists, engineers, entrepreneurs and technicians that are needed to populate a productive innovation system.

In addition to this obvious role, universities can also play a more direct role in the innovation process. Universities can use public funds to develop valuable commercial technology. One of the most critical aspects

1

of any innovation system is whether adequate resources are committed to research and development (R&D). Economic theory posits that for even the most free-market economies, the business sector will always under-invest in R&D.[4] Government investment in research, if properly designed and executed, can help to overcome this problem. The government can act as a collectivizing agent for society as a whole, and fund R&D – frequently through research universities – that may not be justified by private returns, but is capable of generating sizable social returns. In order to capture those social returns, however, universities must develop partnerships with the business sector that allow for the transfer of university-developed R&D into the business sector. These partnerships can range from informal connections (e.g., publication of academic research that is read by the business sector) to formal, contractual mechanisms for transferring technology (e.g., licensing university-developed technology to the business sector and creating start-up companies that are based on academic R&D).[5]

This book will deal primarily with the more direct role played by universities in the innovation process: namely, universities as developers of commercial R&D and the formal, contractual measures for transferring this technology to the business sector so that it can be used to better society. For more than two decades now, China has been working to develop the technology commercialization capacity of its universities. Substantial progress has been made, and university technology commercialization has the potential to become a crown jewel for Chinese economic development at some point in the future. But that time has not yet arrived. This book explores the rise of technology commercialization by Chinese universities and examines why such commercialization efforts have remained on the periphery of China's innovation system. Because modern university technology commercialization strategies are very much 'law-based' strategies – i.e., they are based on well-defined property rights in innovations and the legal transfer of such rights through enforceable contracts – this book's primary focus is to examine whether China's laws, regulations and legal institutions (collectively, the legal system) are sufficient to support such commercialization efforts.

In order to conduct that legal analysis, however, we have also had to conduct a much more holistic analysis of China's overall innovation system. Innovations and technology advances – including university technology commercialization – take place within a complex system that is frequently referred to as an 'innovation system.' Consider the fact that China's legal system has long provided its universities with the ability to transfer the inventions of university researchers to the business sector for commercialization. In 1985, for example, China's State Council issued Provisional Regulations of the State Council on Technology Transfer[6] (the

Provisional Technology Transfer Regulations) that expressly gave Chinese universities, as well as government research institutions (GRIs), the right to manage, transfer and generate income from the work-related inventions of their researchers.[7] Title to the inventions officially remained with the national government, but universities were given critical 'income' rights in the property. These Provisional Technology Transfer Regulations were not overly successful,[8] and have since expired. Why did the Provisional Technology Transfer Regulations fail? Was there something wrong with the laws? Or was the there some other, more fundamental problem in China's innovation system that prevented the Provisional Technology Transfer Regulations from ever having a chance to be successful? If there were more fundamental problems, have they been sufficiently resolved so that China's current technology commercialization efforts have a more realistic chance to succeed? By examining the commercialization efforts of China's universities through the lens of China's overall innovation system, we believe that we have been able to provide much more accurate analysis of the legal system's ability to support technology commercialization and to generate useful insights on the strengths and weaknesses of such commercialization efforts.

1. INNOVATION SYSTEMS

While most policy makers understand the need to develop innovative economies, the best methods for doing that are far from clear. How should policy makers organize their economies to create an environment that consistently creates and integrates technological innovations? The study of innovation systems has developed to try and answer that question.[9] Broadly speaking, innovation system analysis views the process of innovation as a multidimensional one that includes both a variety of *actors* and *institutions* and involves interactions and exchanges between them.[10] In industrialized countries, private businesses, rather than universities, tend to be the dominant innovation system actor.[11] In addition to its obvious role of commercializing technology, the business sector also tends to be the 'dominant locus' for the R&D function.[12] While the business sector is the dominant player, it is by no means the only important player in an innovation system. Universities, primary and secondary schools, GRIs, government officials, venture capital and other sources of finance, and technology consumers, for example, all are important actors in a thriving innovation system (see Figure 1.1, which provides a diagram of the basic actors and activities in a modern innovation system).[13] The ability for these actors to positively influence the innovation process is

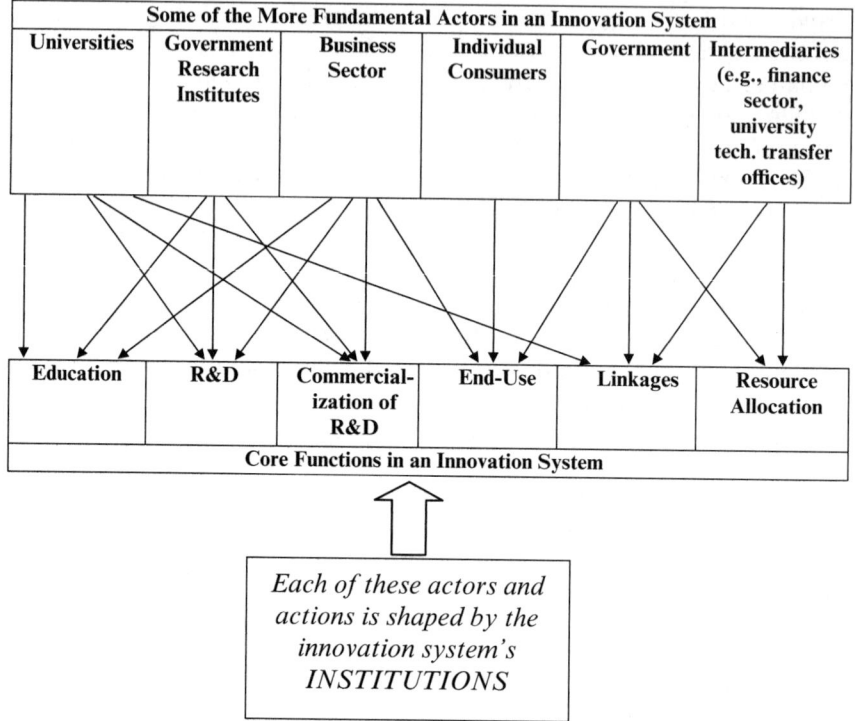

Some of the More Fundamental Actors in an Innovation System					
Universities	Government Research Institutes	Business Sector	Individual Consumers	Government	Intermediaries (e.g., finance sector, university tech. transfer offices)

Education	R&D	Commercial-ization of R&D	End-Use	Linkages	Resource Allocation
Core Functions in an Innovation System					

Each of these actors and actions is shaped by the innovation system's INSTITUTIONS

Note: Figure 1.1 was inspired by a diagram of China's innovation system that was developed by Xielin Liu and Steven White (2001), 'Comparing Innovation Systems: A Framework and Application to China's Transitional Context', 30 *Research Policy*, 1091, 1094.

Figure 1.1   Distribution of actors and activities in a modern innovation system

highly dependent on the various exchanges that take place between them. Innovation tends to be an evolutionary process that is shaped by the connections and interrelationships (e.g., networks) that develop amongst the various actors.[14] Individual actors do not innovate in isolation.[15] Take the example of a high-technology entrepreneur who wishes to develop a new software program. Her ability to innovate will be influenced by a multitude of factors, including the quality of her education, her ability to obtain funding to finance her venture, the technological sophistication of technology consumers in her market, and the willingness of government officials to create an attractive environment for innovation.

Innovation system analysis also focuses on the *institutions* that shape the various actors' behavior and the quality of their exchanges with each other. Economic historian Douglass North[16] provides the standard economist definition for institutions:

> Institutions are the humanly devised constraints that structure human interaction. They are made up of formal constraints (rules, laws, constitutions), informal constraints (norms of behavior, conventions, and self imposed codes of conduct), and their enforcement characteristics. Together they define the incentive structure of societies and specifically economies.[17]

In plain English, institutions are rules created by humans that significantly influence the behavior of humans and organizations. The rules can be formal laws or regulations, but they can just as easily be informal rules of behavior or culture. Institutions are frequently referred to as being akin to the 'rules of a game.' If creating useful technology is viewed as a game, the institutions are the rules for playing that game. They set the rewards for desirable behavior (such as wealth or improved status) and the punishments for deviant behavior (such as financial loss, shame or criminal punishment).

Ironically, having an appreciation of the complex nature of innovation systems makes understanding how they work much simpler. The most innovative economies are those economies with:

- The strongest set of innovation actors;
- The best linkages and coordination between the various actors and their activities; and
- The best institutions (or rules) to motivate the behavior of the various actors.

Analyzing any one of the factors, without an appreciation of how it fits into the overall innovation system will likely generate misleading results. As we explained above, an analysis of whether China's legal system is sufficient to support the commercialization efforts of its research universities is only marginally useful unless placed in the context of China's overall innovation system.

2. THE BAYH–DOLE ACT – THE UNITED STATES PROVIDES THE LEADING EXAMPLE

The United States provides the clearest, and most mature, example of a successful environment for university/industry partnerships and the resulting technology commercialization benefits. While U.S. universities

Table 1.1   Summary data from AUTM's annual survey of U.S. universities (2005–2008)

	2005	2006	2007	2008
1. Patents (includes U.S. universities and U.S. hospitals & research institutes)				
Number of respondents	191	189	193	191*
Invention disclosures received	17 382	18 874	19 827	20 155
New patent applications filed	10 270	11 622	11 797	12 072
U.S. patents issued	3278	3255	3622	3280
2. Licensing Activity (U.S. universities only)				
Number of respondents	158	161	161	159
No. licenses and options executed	4201	4192	4419	4376
3. Commercial Products				
New products introduced into the market based on academic inventions	527	697	686	648
4. Start-up Activity				
University spin-offs created	628	553	555	595

Note: * In 2008, there were 191 respondents for the 'new patent applications field' and 'U.S. patents issued' statistics, but only 189 respondents for 'invention disclosures received.'

Source: The Association of University Technology Managers (AUTM), *U.S. Licensing Activity Surveys – Survey Summaries – FY 2005–FY 2008.*

are typically tasked with studying basic, or foundational, science, rather than seeking to resolve a specific commercial problem, those foundational efforts can lead to valuable commercial technology. The study of foundational physics, for example, helped lead to the modern computer.[18] Each year, U.S. universities generate a considerable amount of cutting-edge technology and efficiently disseminate that technology to the public through the business sector. The Association of University Technology Managers (AUTM) annually conducts a survey of U.S. universities to measure their ability to move technology into the marketplace. Table 1.1 provides select summary data from AUTM's annual surveys from 2005 through 2008.

Many credit a legal innovation that took place in 1980 with the United States' dramatic success in university technology commercialization. The U.S. Congress enacted the Bayh–Dole Act of 1980[19] in December of 1980 to strengthen the rights of universities over intellectual property (IP) arising from federally-funded research. An editorial that appeared in the December 2002 issue of the *Economist*, for example, provided the following enthusiastic recommendation of the Bayh–Dole Act:

> Possibly the most inspired piece of legislation to be enacted in America over the past half-century was the Bayh–Dole Act of 1980. Together with amendments in 1984 and augmentation in 1986, this unlocked all the inventions and discoveries that had been made in laboratories throughout the United States with the help of taxpayers' money. More than anything this single policy measure helped to reverse America's precipitous slide into industrial irrelevance.[20]

Prior to the Bayh–Dole Act, the federal government retained the IP rights to federally-funded inventions. The theory behind the pre-Bayh–Dole approach was that the federal government, on behalf of taxpayers, should be entitled to capture the economic rewards of the research that it funds. This capture was accomplished by the federal government licensing the resulting IP to industry, in most instances, on a 'nonexclusive' basis.[21] Apparently, these nonexclusive licenses were not sufficiently attractive to industry. A 1978 report by the Federal Council for Science and Technology found that the federal government held title to approximately 27 500 patents, but only 4.5% of these patents had ever been licensed to industry for commercial development.[22] The Bayh–Dole Act helped to remedy this situation by altering the property rights of federally-funded inventions. Instead of granting those property rights exclusively to the federal government, the Bayh–Dole Act provides universities and small businesses with the right to keep the title in their federally-funded inventions, subject to certain encumbrances and rights retained by the government.

The Bayh–Dole Act provided a number of tangible benefits to the U.S. innovation system:

- From the universities' perspective, the Bayh–Dole Act provided a clear incentive to develop and commercialize valuable technology by establishing strong, well-defined IP rights.
- From the business sector's perspective, research universities became a much richer source of technology – '[i]t turned every research university into a potential research and development partner with whom firms could leverage their research funds and negotiate rights to the innovations created in the partnership.'[23]
- From the government's perspective, it helped to 'mov[e] patents off the shelves of the federal bureaucracy and into the marketplace.'[24] It helped to generate the social benefits that taxpayer-funded research is meant to generate by moving technology into the commercial sector where it could generate economic development. While not an

explicit goal for Congress at the time the Act was passed, the Bayh–Dole Act also provided a market-based mechanism for funding the most effective research universities as these top research universities have been able to generate significant revenues from their technology transfer activities.

3. CHINA'S 'BAYH–DOLE' EFFORTS – THE NEED FOR A HOLISTIC ANALYSIS

China has tried to emulate the United States' Bayh–Dole approach to university technology commercialization. Following the failed Provisional Technology Transfer Regulations, China has experimented with various policies, regulations and laws to provide universities with IP rights to government-funded inventions with the hope of encouraging the universities to engage in commercial dissemination strategies. The Law on Science and Technology Progress (2007) is China's most recent effort on this front and shares many similarities with the Bayh–Dole Act. Chapter 6 will provide a detailed analysis of China's overall collection of technology commercialization policies, regulation and laws, which we will refer to collectively as 'China's Bayh–Dole efforts.'

Since university technology commercialization takes place within a country's overall innovation system, implementing legislation that provides universities with IP rights to government-funded inventions is only one of many important factors that influence the successful development of that partnership. Independent of the Bayh–Dole Act, the United States has a vibrant innovation system that includes, among other things:

- Strong R&D capacity throughout its university system that is grounded in a highly-competitive research culture. Researchers must go through highly-competitive processes to get university research positions, to obtain research funding and to disseminate the results of their research through peer-reviewed publications.
- A vibrant high technology business sector that is skilled at absorbing new technology and transforming it into new commercial products and services.
- A history of cooperative relationships between the university and business sectors.
- A federal government that appreciates technology's importance to economic development and provides strong financial support for university R&D.
- The presence of competent actors and strong institutions to support

a vibrant market for technology (e.g., venture capitalists and other sources of entrepreneurial finance, competent IP lawyers and IP valuation experts, and information institutions to support the full and accurate information that is needed to encourage efficient exchanges).

- Cultural respect for IP and its ownership.
- A legal system that supports market transactions, generally, and IP transactions, specifically. For example, the U.S. legal system is generally IP-friendly, supports the various actors that function within the innovation system (e.g., allows for efficient business association structures, regulates securities transactions, provides a reasonable corporate governance system), and possesses a judicial and administrative process system that competently enforces the various property and contract rights that form the basis of technology exchanges.
- University administrators that build institutions within the university to support the commercialization process and create a technology commercialization culture.

In the United States, the Bayh–Dole Act was asked to do very little. With the possible exception of university IP management offices and the venture capital industry (both of which developed substantially post-Bayh–Dole), most of the core elements of its vibrant innovation system were already in place. While this is a bit of an over-simplification, Bayh–Dole was only called on to improve the innovation incentive structure by shifting a property right from one holder to a more logical holder. For economic actors to be incentivized to invest in the creation and improvement of property on a consistent basis, they must believe that they will be able to collect the returns from those efforts.[25] By shifting the IP rights to universities, the key economic actor (i.e., universities) now had the security to make the necessary resource commitments for developing and disseminating commercially viable R&D.

For China, however, analyzing the impact of its Bayh–Dole efforts on academic/industry partnerships requires a much deeper analysis. China's Bayh–Dole efforts must be viewed in context. Whether it will be ultimately successful is very much dependent on China's overall innovation environment – not just on whether Chinese policymakers have adopted the right technology commercialization laws. With China, one cannot simply assume a vibrant, well-functioning market-based innovation system, the only concern of which is the clarity of IP rights in government-funded academic inventions. As a result, this book will cover a wide variety of matters.

Part I of this book examines the quality of operations of China's overall innovation system.

Chapter 2 – Developing a Market-Based Innovation System: employing a Bayh–Dole approach to encourage university innovation and technology transfer is predicated on there being a market-based innovation environment. Chapter 2 chronicles the broad changes that have taken place in China's innovation system during China's three decades of economic reform. To place the reforms in context, Chapter 2 also provides an overview of China's plan-based innovation system during the Mao era and explains how legacies from that era continue to impact China's current innovation system.

Chapter 3 – A Snapshot of China's Current Innovation System: China's innovation system made substantial strides throughout the 1990s and 2000s and has experienced some of the strongest institutional growth in China during that period. Chapter 3 examines the current roles and capacities of China's university and business sectors and their ability to meaningfully collaborate through a Bayh–Dole strategy.

Part II of this book analyses the legal and policy environment for commercializing university technology in China. A country's national policies play a critical role in creating an environment that either encourages or discourages the formal flow of commercial technology from universities to the business sector. In addition to having a market-based innovation system, there are five critically important policies that a government needs to consider:[26] (1) does the country's legal system support market-based transactions in general?; (2) does the country's legal system specifically protect IP on a sufficient level?; (3) who owns the IP rights to government-funded, university inventions, and what rights are afforded to the owner?; (4) does the government dedicate sufficient public funds to university research, and are those funds allocated in an efficient manner?; (5) does the government provide a sufficiently supportive environment for innovative, technology-based companies so that universities have a sufficiently strong business sector with which to partner? Part II will examine each of these issues in detail.

Chapter Four – Developing a Legal System to Support Market-Based Transactions: a Bayh–Dole strategy seeks to use the law to encourage technology-based economic development. By shifting property rights from the government to universities, Bayh–Dole employs a market-based legal strategy to encourage the creation of new technologies and to move such new technologies into the marketplace where they can best benefit the public. In developed countries, little thought is given to whether a country's legal system supports market-based transactions, as the legal system has likely performed such a role for centuries. In 1978, at the start of China's reform period, the legal infrastructure that people take for granted in most countries was non-existent in China. At that time, there were almost no lawyers in China, there was only a minimal amount of civil law on the books and the existing court system was not capable of handling commercial disputes. Chapter 4 explores the gradual, but fundamental, change in the economic role of law in China since that time. While China's legal system was a largely irrelevant institution at the outset of the reform era, Chapter 4 chronicles the progress that has been made in China's legal system and explains why it is now strong enough to support a market-based, Bayh–Dole strategy.

Chapter 5 – China's Intellectual Property Regime has Come of Age: IP protection provides the foundation for a technology transfer system. At the most general level, IP protection provides an economic incentive for individuals and firms, including universities, to invest in R&D.[27] At a more specific level, IP protection provides the exchangeable property rights that are a pre-requisite to a Bayh–Dole approach to technology transfer. Chapter 5 examines China's IP protection system and will explain how China has implemented a modern IP regime that is more than sufficient to support a Bayh–Dole approach to technology transfer.

Chapter 6 – China's Bayh–Dole System: for more than two decades, China has worked to create a viable market for technology created by its publicly-funded research institutions. Chapter 6 explores China's legal framework for commercializing university-developed technology, including the economic incentives that result from that framework. Chapter 6 also examines the positive results that have come from those efforts.

Chapter 7 – Planning to be an Innovative Nation – China's National S&T Plan and its Impact on China's Bayh–Dole Efforts: two factors that will play a prominent role in whether technology transfer will continue to grow in China are: (1) the willingness of the government to continue to fund university R&D at sufficiently high levels and the efficiency of that funding effort; and (2) the R&D proficiency of China's business sector. Chapter 7 examines China's recently adopted national S&T policy, which provides explicit and positive guidance on both of these issues.

Chapter 8 – China's Emerging Venture Capital Industry: Chapter 8 continues with the issue of the business sector's R&D strength in China, but will look at the issue from a different angle. Chapter 8 focuses on one particular element of China's business sector that tends to be a very natural technology transfer partner for universities in developed countries: high-technology start-up firms (start-ups). The creation and survival of start-ups is highly dependent on their ability to procure capital, with venture capital firms serving as the most skilled and efficient furnishers of such capital in the developed world. China has been working for more than a decade to develop a modern venture capital industry that is capable of promoting the growth and creation of a vibrant start-up industry. Chapter 8 examines the efficacy of these policy efforts and what they mean for China's start-up industry future.

Part III will look towards the future of technology commercialization in China. What is needed to make it better?

Chapter 9 – Increasing the Technology Commercialization Capacity of Chinese Universities:　　Chapter 9 considers what comes next for China's Bayh–Dole efforts and offers a number of concrete recommendations for improving China's university technology commercialization system.

We hope you enjoy this look at the increasingly important technology commercialization role of Chinese universities. Studying the birth and development of a modern innovation system in China – including the development of its various institutions – to the point that a Bayh–Dole approach to technology commercialization has been able to take shape has been nothing less than fascinating. A lot of work remains to be done in order for Chinese universities to generate and commercialize valuable innovations that can consistently contribute to China's economic growth. However, based on China's remarkable progress over the last three decades in developing a market-based economy, rejuvenating its university sector, creating a modern legal and IP system, promoting a modern venture capital industry, and encouraging stronger linkages between the various innovation system actors, it strikes us as very reasonable to believe that China and its universities will be up to the challenge.

We would like to make one final note before starting our analysis. While this book focuses primarily on university/industry technology commercialization, much of our analysis applies equally to technology commercialization for GRIs. As a result, we will provide an overview of GRIs in China and will try to integrate them into our analysis when possible.

PART I

China's innovation system: Mao, markets and the growing prominence of Chinese universities

2. Developing a market-based innovation system

> The people are happy and we have captured the attention of the world.
> – Deng Xiaoping (Jan. 1992)[1]

Employing a Bayh–Dole approach to encourage university innovation and technology transfer is predicated on there being a market-based innovation environment. This chapter chronicles the broad changes that have taken place in China's innovation system during China's three decades of economic reform that have transformed China's former 'plan-based' innovation system into one that is largely market based. Despite this progress, legacies from the Mao era continue to impact China and must be accounted for in any innovation system analysis for the country.

Before launching into a review of China's market-based transformation, however, it is useful to begin with a brief overview of markets, their function in a society, and how a Bayh–Dole strategy fits into such a market-based system.

1. BRIEF OVERVIEW OF 'MARKETS' AND HOW THEY RELATE TO BAYH–DOLE

Scarcity provides the foundation for the field of economics. Most of the goods and services that we consume each day (e.g., food, clothing, and consumer goods) are scarce – meaning that the demand for such goods or services is greater than the supply. Take the simple example of *time*. Time is a scarce resource for humans relative to the almost infinite ways that we could spend our time in a given day. Because time is finite, while the possible ways of spending it are limitless (or at least almost limitless), we must make choices on how we spend our time. We must decide how to allocate our time to accomplish a variety of tasks in any given day – e.g., how much time we should spend working, sleeping, eating, playing, learning, etc. The primary function of an economy – whether a market-based economy or a Soviet-style planned economy – is to address this scarcity problem. More specifically, an economy needs to allocate scarce resources

15

for society – preferably in a way that creates the most value for society. To do that, an economy must make three fundamental decisions:[2]

- *Production:* what goods and services should be produced, and how much of each should be produced?
- *Distribution:* how much of each good or service should be allocated to any given actor in the economy?
- *Financing:* what amount of resources should be provided to the various actors to help them to produce, buy or sell goods or services?

In a planned economy, these various decisions are made by the government through some type of planning function. The planning function decides what is to be produced (and how much), how the produced goods or services are to be distributed, and how much to invest in production.[3]

In a market-based economy, the three fundamental decisions are not centralized under some planning authority, but instead are distributed amongst various actors in the economy who are then required to compete against each other through a constant process of exchange. One party exchanges the scarce resources that it owns for scarce resources owned by another party. This competitive exchange process helps to resolve each of the three fundamental decisions.

- *Production:* producers are given general freedom to produce what they want, but must satisfy consumer demand if they hope to generate sales. Consumers, therefore, drive the decision of what, and how much, to produce through their purchase decisions. In an effort to generate the most consumer sales, producers must compete with each other to provide the most desirable products and services to consumers at the best prices.
- *Distribution:* producers are not the only parties required to compete. The decision on what quantity of goods or services to distribute to any given consumer is also answered through competitive exchange. Consumers must compete to obtain scarce resources, which they typically do by offering to pay the highest price for the goods or services.
- *Financing:* a competitive market process also leads to more efficient allocation of the investment capital that finances the various transactions. Sources of capital seek to allocate their investment funds to projects that are likely to generate the greatest returns, while recipients of capital seek to obtain the lowest cost capital.

Exchange is the driving principle of a market-based system. Through a desire to promote their own self interest, buyers and sellers *exchange* scarce resources – with each party seeking to obtain a more valuable resource than it surrendered – which helps to move scarce resources to higher valued uses.[4]

A Bayh–Dole-type approach to encouraging university innovation and commercialization aims to capitalize on the basic exchange-based incentive structures that markets promote. For economic actors to be incentivized to invest in the creation and improvement of property on a consistent basis, they must believe that they will be able to collect the returns generated by those efforts.[5] The basic function of the Bayh–Dole-type approach is to place exchangeable property rights in the hands of the actor (e.g., universities) that is in the best position to develop that property and to place it into the flow of commerce. Assuming that a market exists to exchange these rights for monetary or other valuable resources, universities typically collect the returns from their commercially-viable R&D efforts by patenting their inventions and either licensing the patented technology to the business sector or establishing start-ups that control the patented technology.

2. THE MAO-ERA PLAN-BASED INNOVATION SYSTEM (1949–1978)

China operated a plan-based innovation system for most of the second-half of the twentieth century, and only began to make the transition to a more market-based innovation system in the mid-1980s. After the establishment of the People's Republic of China in 1949, China moved to a Soviet-style, centrally-planned economy. Included in this central-planning transition was China's innovation system.

2.1 The Principles of Functional Specialization

In accordance with Soviet principles of functional specialization, China's primary innovation activities were separated and formally distributed amongst a limited set of technology actors (see Figure 2.1). Specific boundaries were established round each activity, and actors were encouraged to remain within their narrowly defined boundaries.[6] Under the functional-specialization system, China organized the innovation process into six fundamental activities:

1. *Education and training of S&T personnel*: schools (including universities, vocational schools and technical schools) were responsible for educating the various S&T researchers and workers.
2. *Scientific research*: GRIs were established to conduct scientific

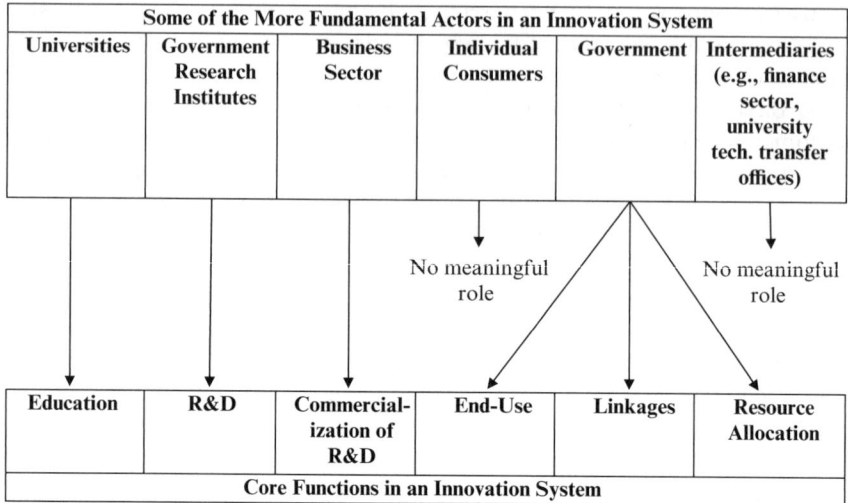

Some of the More Fundamental Actors in an Innovation System					
Universities	**Government Research Institutes**	**Business Sector**	**Individual Consumers**	**Government**	**Intermediaries (e.g., finance sector, university tech. transfer offices)**

No meaningful role	No meaningful role

Education	**R&D**	**Commercial-ization of R&D**	**End-Use**	**Linkages**	**Resource Allocation**
Core Functions in an Innovation System					

Note: Figure 2.1 was inspired by a diagram of China's innovation system that was developed by Xielin Liu and Steven White (2001), 'Comparing Innovation Systems: A Framework and Application to China's Transitional Context', 30 *Research Policy* 1091, 1094.

Figure 2.1   Distribution of actors and activities under China's Mao-era plan-based innovation system

research. The biggest and most successful of the GRIs was the Chinese Academy of Sciences (CAS), which continues to operate today. In the functional-specialization system, the CAS primarily focused on basic research. Smaller industrial and local GRIs focused on applied R&D and developmental activities.[7] While there were a few exceptions,[8] businesses and universities did not play very significant roles in scientific research.[9]

3. *Commercialization of the research*: the transformation of scientific research into useful products and services was conducted by state-owned manufacturing firms (factories or production enterprises).[10]

4. *End-use*: somebody needs to consume the commercial products or services. State-owned enterprises were the largest consumers in China during the functional-specialization era. Individual consumers also served as end-users, but in no way had the influential impact that one sees in a market-based system.

5. *Linkages*: a variety of governmental actors at both the central level (e.g., the State Planning Commission and the State Science and

Technology Commission) and the local level worked to coordinate the various innovation activities. In some cases, there were multiple government actors (commissions, ministries or industrial bureaus) with overlapping oversight of the same innovation activity.[11]

6. *Resource allocation*: government actors dominated the myriad of resource allocation decisions that occur within an innovation system. Government actors decided what research to conduct, and the level of resources that would be committed to any given S&T project. Government actors also decided what the manufacturing firms should do with particular research and established a narrow range of choices that would be offered to the end users. Finally, government actors established the education agenda for the school system.

From the perspective of the basic technology actors (other than the government), China's 'functional specialization' approach tended to leave them with inappropriately narrow roles in the innovation process or eliminated their role altogether. GRIs, for example, were limited to conducting research, but they had little say in the direction of that research or how their resources should be allocated. The business sector's role was even more limited. While a few of the larger factories/production enterprises had in-house research facilities, most did not conduct any R&D.[12] For the most part, businesses were limited to manufacturing and distributing the resulting technology products. Schools performed their traditional education function, but had little say in curricular development and, with a few exceptions, did not share in the research function.

Compartmentalizing the innovation process into such narrow boundaries caused some very serious problems for China's innovation system that continue to be felt today.

2.1.1 Eliminated entire categories of important technology actors

While GRIs, the business sector and schools had very limited roles, entire categories of technology actors were effectively eliminated from the innovation process. Private technology consumers (both business technology users and individual consumers) had no meaningful role in the plan-based innovation system, since technology exchange and resource allocation decisions were not based on purchasers' choices. There was no internal market for technology in China. Instead, the government directed the transfer of technology and the allocation of resources throughout the innovation system. The lack of a technology market also eliminated private technology intermediaries from the innovation process. Banks, venture capital firms and other sources of capital, for example, were rendered irrelevant since such resource allocation decisions were the responsibility of the government. Technology transfer professionals – e.g., actors that seek to facilitate technology exchanges – were also rendered irrelevant. Without a market for technology, technology transfer professionals simply did not have a role.

2.1.2 Eliminated the naturally forming linkages between the various actors and functions

Naturally forming linkages between the various actors and functions in China's innovation system were eliminated and replaced by linkages that were dictated by the government. Rather than encourage a self-generating 'information loop' whereby the various actors provide each other with information about their needs and capabilities, the plan-based system forced the various actors to interact with each other based on top-down allocation decisions from government bureaucrats.[13] Highly ineffective communication channels, which substantially reduced the efficiency of the innovation process, were the result.

One of the more concrete examples of this 'de-linkage' was a formal, concrete separation of research from industry. From the business sector's perspective, it could not directly communicate its technology needs (or the technology needs of its customers) to researchers – which resulted in researchers taking their directives from government policymakers rather than business leaders (discussed below in this chapter – section 2.1.3). The business sector was also unable to communicate its ability to absorb technological advances to researchers, so there was a strong likelihood that technological advances would not mesh with the business sector's actual capacity to make productive use of the advances. Researchers fared no better from their separation from industry, as they could not actively market their breakthroughs to technology users, nor could they seek feedback on the usefulness of their breakthroughs.

The separation of research from industry is a serious problem that has yet to be fully resolved in China. Even 30 years after the initiation of market-oriented reforms, China's business sector has not yet fully embraced the value of conducting private R&D. Chapter 3 will examine the relatively weak nature of the R&D capabilities of China's business sector. Three decades of functional specialization is almost certainly a major reason for the failure of China's business sector to fully embrace privately-funded R&D.[14]

2.1.3 Lack of commercially-focused research

During the central-planning era, the decisions to pursue particular R&D projects (and the funding decisions for those projects) were made by the government, rather than driven by commercial needs. Two of the most important government actors were the State Development Planning Commission

(SDPC) and the State Science and Technology Commission (SSTC). The SDPC developed the Five-Year Plans that established national objectives and allocated resources for each of China's administrative and production units. The SSTC then had specific oversight over science and technology activities.[15] During much of this period, China struggled with substantial concerns about its national security as well as extreme scarcity of food and other basic resources. These compelling concerns dominated China's research agenda rather than the development of commercially-viable technology. Without consistent input on commercial needs, the focus of central planners understandably tended to be towards these military/livelihood issues, but the result was a technologically depressed business sector. China's greatest technological achievements during the central planning era were, not surprisingly, high priority military projects.[16]

2.1.4 Eliminated many of the most important innovation institutions

One of the most obvious institutional casualties of China's plan-based innovation system was the legal system. During the planned-economy era, the legal system's role in the innovation system was rendered largely irrelevant. Allocation and ownership decisions were made by the government, rather than privately negotiated contracts surrounding the ownership of property. As a result, the various legal institutions that typically support an innovation system held little importance and were not developed. At the start of the reform era in 1978, China simply did not have working institutions for property law (including intellectual property rights), contract law, company law, banking law, securities regulation, competition law, or bankruptcy law, to name just a few. Developing such legal systems almost from scratch is not an overnight process and should be measured in decades, not years. China has been working diligently to develop its legal system, which will be examined in Part II.

The law is just one of many institutions that were diminished or eliminated during the planned-economy era. The institutions of education and science were also greatly damaged. China's financial institutions, which were rendered largely useless by the planned-economy apparatus, were another casualty of the planned-economy era. Just as with developing a well-functioning legal system, it takes time to develop these various innovation institutions.

On a much more general level, the planned economy greatly weakened the typical incentives for innovation.[17] In a market-oriented environment, gaining a technological advantage over competitors is one of the most powerful incentives for innovation. Frequently referred to as involving a process of 'creative destruction', participants in more market-based economies are motivated to seek innovations to beat competitors by 'creating' new products, services, or processes for doing business more efficiently.[18] Innovative companies flourish, while less efficient ones are destroyed. In China, this creative destruction process was replaced by government over-sight. Chinese researchers Xielin Liu and Steven White describe China's plan-based incentive problem as follows:

> There was neither market competition, profit, nor other operational efficiency-based criterion for performance. Nor were there any other institutional mechanisms absent government direction that encouraged primary actors (or the individuals within them) to improve upon the activities in their mandate, such as investing in technology development or adoption, or upgrade existing technology. The central government secondary actors claimed both authority and responsibility for such initiatives. Each primary actor's participation was limited to bargaining with the [officials] within the central government over resources and output goals for their organization, whether R&D output, manufacturing output, students, or whatever was within the organization's functional domain.[19]

Not surprisingly, the weak incentive structure translated into a weak innovation record during the Mao era, which caused China to start from a much lower development point at the outset of the reform era. For example, China imported the design and the production lines for the Liberation Truck from the Soviet Union in the early 1950s, but failed to make any significant technological advances to the truck's design or production during the 40 years in which China mass-produced it.[20] Unlike most of the other institutional problems, however, this basic market-oriented incentive problem can be corrected rather quickly, and in many ways has already been corrected.

2.2   The Cultural Revolution

As bad as the problems were during the planned-economy era, China's catastrophic 'Cultural Revolution' made them worse.[21] The Cultural Revolution began in late 1965 and sent China into a 10-year period of chaos. China's youth (organized into Red Guard groups) was encouraged to revolt against the intelligentsia and others deemed to be developing into a new privileged class. The Cultural Revolution substantially disrupted China's economy, as well as its R&D and educational systems. With the exception of certain military research projects, China's R&D and university systems failed to function normally for much of the 1970s.[22]

The ability of China's universities to develop China's human capital (in terms of both highly-trained S&T personnel and S&T researchers) was very seriously compromised during those ten years. In addition, scientists and engineers were denounced and dismissed from their positions.[23] It is difficult to measure the long-term impact of such a sustained period of educational and R&D disruption. In a recent book on China's economic transformation, Professor Gregory Chow theorized that one reason for the success of China's market-oriented agricultural reforms compared to Russia's agricultural reforms was that the disruption in knowledge transfer for China's farmers was relatively short-lived.

> China's reform in the agricultural sector succeeded because Chinese farmers in 1978 still remembered private farming – the commune system was introduced in China only in 1958. Collectivization of agriculture in the Soviet Union took place in the early 1930s and Russian farmers in the early 1990s did not know how to operate as private farmers. It was difficult to privatize the agricultural sector in Russia.[24]

In Russia, the knowledge transfer was disrupted for such a long period of time that Russian farmers had to almost 'start from scratch' in learning how to operate a private farm. The Cultural Revolution's disruption to China's innovation system may have been similar. In many ways, China's universities and primary and secondary schools had to 'start from scratch' when Deng Xiaoping ended the Cultural Revolution in 1976, which provides a partial explanation for why China's domestic innovation rates were incredibly low through the mid-1990s, at which point they began to show some progress. China was operating under a scientist/professor gap from 1978 through the mid-1990s that is only recently being filled.

2.3 China's Innovation System was at a Very Low Point at the Outset of the Reform Era

All told, China's innovation system began the reform era in a rather pitiable state. While the point is rather obvious, we find in our discussions with others about China's innovation system that there is a tendency to forget how much progress China's innovation system has made since 1978. In 1978, China had a poorly functioning system of GRIs and the business sector provided almost no assistance to the R&D process. China also suffered from a weakened education system that included a largely non-functioning university system. To top things off, China was one of the poorest countries in the world at that time. In 1978, China did not have even the most basic ingredients for a vibrant innovation system. It did not have a sufficiently competent base of researchers or technology workers and it lacked the financial resources to commit to the S&T effort.

The low starting point of China's innovation system had a direct impact on the development of China's innovation institutions, including its legal system. In 1978, when China lacked the basic ingredients for a vibrant innovation system, there was little demand for innovation institutions, and therefore little incentive to develop them in earnest. By the 1990s, China was developing a competent base of S&T human capital and had truly meaningful resources to dedicate to S&T. It was at this point that the need for innovation institutions became much more tangible and began to grow. Chapters 4 and 5 will explain how China's innovation-supporting legal system truly began to develop in the early 1990s. Few would argue that developing innovation institutions almost from scratch is a long process that is best measured in decades, rather than years. When measuring the progress of China's institutions, it is probably more accurate to view the early 1990s as the true starting point.

3. TAKING THE INITIAL STEPS TO REVIVE CHINA'S INNOVATION SYSTEM (1978–1984)

In late 1978, under the leadership of Deng Xiaoping, China began its long journey toward a market-based economy with a series of economic reforms. It bears mentioning that most of these early reforms were not market-based in nature. The early innovation system reforms, for example, were a continuation of China's plan-based model that sought to restore China's innovation system to its pre-Cultural Revolution state (i.e., restore an uncorrupted Soviet model).[25] China's S&T weakness was not thought to stem from the plan-based innovation system, but instead was blamed on the chaos of the Cultural Revolution. Ironically, China's efforts to return to its 'plan-based roots' proved instrumental for the later market-based reforms to the innovation system.

The National Science and Technology Program (NSTP), which ran from 1978 to 1985, is a classic example of these early plan-based reforms. The NSTP sought to reinvigorate China's GRIs and increase the focus on civilian-based R&D.[26] The NSTP had three major focuses:

- Create new GRIs and increase government spending on S&T (e.g., government expenditures on S&T increased from 4.8% of the government budget in 1978 to 5.3% in 1980);[27]
- Improve the management of individual GRIs (e.g., by inserting experts as GRI directors in place of political appointees);[28] and
- Encourage enterprises to create their own R&D departments.[29]

In typical plan-based style, the NSTP identified eight S&T fields as national priorities (agriculture, energy resources, materials, computing, lasers, space science and technology, high-energy physics, and genetic engineering) in which China's GRIs were to attain a leading status.[30]

While China's S&T system was improving, at least compared to its operation during the Cultural Revolution, the Chinese government recognized that the efforts of its GRIs were not well integrated with China's overall economic efforts.[31] The level of commercially-useful technology that resulted from GRIs remained disappointing.[32] The government responded with a series of reforms that were meant to improve the coordination of China's R&D efforts with its economic activity. In 1982, for example, the Ministry of Science and Technology enacted the Key Technology R&D Program[33] to concentrate China's S&T resources more efficiently on China's most pressing economic concerns and social development. Among other things, the Key Technology R&D Program sought to accelerate the development of key technologies to support agriculture and traditional industrial activities and also to promote new, high technology industries.[34]

By the mid-1980s, China's innovation system was facing a new set of challenges. China had substantially grown its S&T assets, but it had done so in a way that was extremely inefficient and wasteful. By 1985, there were close to 5000 GRIs operating at the national level, with an additional 3000 operating at the local level, and more than 300 000 scientists and engineers working at these various GRIs.[35] Not surprisingly, the various GRIs were incapable of coordinating their R&D activities in anything close to an efficient manner, which frequently led to duplicative efforts, and has been referred to by one study as 'chaotic plurality.'[36] The GRIs also had ineffective relationships with the business sector. A Canadian mission of S&T experts that traveled to China in 1995 described the operation of China's mid-1980s GRIs as follows:

> It appears that the majority of the work done at these institutes was not R&D but was much more oriented to supporting SOEs [state-owned enterprises] that were weak in development capability and had little incentive to innovate. Paradoxically, this assisted such institutes in the reform process because, once given more freedom of operation, they were able to transform themselves into profitable businesses.[37]

While the initial reform efforts for China's innovation system were not overly successful, they did plant the seeds for future market-based reforms. To begin with, they clearly demonstrated to China's government the difficulty (and probably the impossibility) of trying to centrally plan such a dynamic and complex system as an innovation system. By the mid-1980s, it was clear that the failing of China's plan-based innovation could not be blamed entirely on the chaos of the Cultural Revolution. The Cultural Revolution definitely made things worse, but the plan-based system itself was now being exposed as problematic.

The early reforms also played a more subtle role in the market-based movement. The apparent loosening of regulations during that period motivated a number of scientists to quietly challenge the plan-based system and become successful entrepreneurs, which helped to prove the value of a more market-based approach to innovation. Though this more subtle role of the early reforms is frequently missed, it could be one of the more important drivers of the movement. During the early 1980s, for example, a few scientists from the CAS formed spin-off companies based on CAS-developed technology.[38] These spin-offs were not conducted pursuant to a formal reform effort (e.g., the Torch Program), but instead were a true market-based reaction by the scientists. In 1980, the CAS Institute of Physics created Beijing Huaxia Guigu Information System Corporation Ltd., which was China's first privately-owned high-technology company.[39] Other CAS spin-offs followed, which created a convenient vehicle for converting CAS technology into commercial products. Some of the more famous of these early CAS spin-offs included:[40]

- The four 'pioneer corporations' of Zhongguancun Science Park (1983–1984): Kehai Corporation, Jinghai Corporation, Stone Group and Xintong Corporation.[41] Stone Group, for example, began by producing computer printers and Chinese–English electronic typewriters. Stone Group was also one of the early Chinese companies to export its technology abroad.[42]
- New Technology Company (1984): New Technology Company was the predecessor of the Lenovo Group.

The success of these early CAS spin-offs helped pave the way for China's future reforms of its innovation system (e.g., the Torch Program in 1988). In addition to showing the value that start-ups can play in commercializing technology, the early CAS spin-offs also provided a working model for the development of China's high-technology Development Zones[43] (discussed below in this chapter – section 4.3). The initial CAS spin-offs tended to cluster informally in Zhongguancun (in Beijing's Haidian district), eventually earning this area the nickname of 'Electronics Street.' The State Council approved Zhongguancun as China's first Development Zone in 1988.[44]

4.   INITIATING MARKET-BASED REFORMS AND INCREASING THE COMMERCIALIZATION OF GOVERNMENT-FUNDED RESEARCH (1984–1992)

Recognizing the innovation system's continuing weaknesses, China's leadership began the more fundamental task of implementing market-based reforms to the system during the mid-1980s. Table 2.1 provides a summary of some of the more critical S&T reforms from this period:

These reform efforts, among others, helped to advance China's innovation system on a number of fronts:

- Encouraged R&D efforts at universities, which were 'barely existent in the pre-reform era.'[45]
- Encouraged the development of horizontal, market-based links between GRIs, research universities and the business sector.[46]
- Increased China's inventory of technology assets by importing technology, which involved a combination of 'acquiring foreign technology, attracting foreign investment, and sending students abroad for training.'[47]
- Decreased the role of administrative fiat in allocating S&T resources, and introduced market-oriented mechanisms as a replacement.[48]
- Expanded and diversified the pool of Chinese individuals that could eventually participate in China's innovation system.
- Began to develop the formal institutions (e.g., a viable legal system) needed to support a market-based innovation system.

Collectively, the reform efforts from 1985–1992 began the overhaul of huge segments of China's innovation system. China's education system, its legal system, its GRIs and its business sector were all the subject of fundamental, market-based reforms that originated during this 1985–1992 period. The education system reforms will be discussed in Chapter 3 and the legal reforms will be discussed in Chapters 4 and 5. The remainder of this chapter will concentrate on China's attempts during this period to (1) create a 'technology market', (2) merge its GRIs directly into business enterprises and (3) develop a Chinese start-up industry.

4.1 Attempting to Create a Technology Market

Through the early 1980s, China's innovation system continued to operate largely under the functional-specialization theory – with universities playing only a limited, but growing, role in China's national research system.[49] GRIs continued to form the nucleus of China's innovation system and, as such, were the logical starting point to initiate market-based reforms. With so much of China's S&T resources being committed to GRIs, something had to be done to realize proper value from that commitment. GRIs needed to be more responsive to the needs of China's economy – in terms both of producing useful technology and transferring that technology to commercial vehicles that could use the technology

Table 2.1　Select early market-based technology reforms (1984–1992)

Year	Reform Effort	Description of Reform Effort
1984	Patent Law enacted	The National People's Congress approved the Patent Law which established the basic legal mechanism for the subsequent efforts to create a 'technology market' in China.
1985	Decision on Reform of the S&T Management System	Attempted to create a 'technology market.' While the technology market effort failed, it marked the beginning of a series of concerted efforts to expressly introduce market-based principles to China's innovation system.
1986	'863' Plan	High-technology research development plan that identified and funded certain key, frontier technologies. The 863 Plan allocated roughly 5 billion RMB between 1986 and 2000 in areas such as biotechnology, information technology, energy, robotics, new materials, space, and lasers.[1] The 863 Plan 'introduced the concept of peer review for the first time in technology research and focused primarily (but not entirely) on civilian objectives.'[2]
1987	Stipulations of the State Council for Furthering the Reform of the S&T Management System	Encouraged GRIs to merge into existing business enterprises. Only a few GRIs (out of the thousands that existed at the time) responded to the merger encouragements. By 1988, the government officially admitted that the program was not successful.[3]
1987	Law of Technological Contracts	Established the legal mechanism for exchanging technology rights.
1988	Torch Program	Program to stimulate the creation of high-technology start-up companies.
1988	Zhongguancun	First high-technology zone was established in Beijing.
1990	Dissemination Plan	Designed to promote and improve the dissemination of technology developed by GRIs.[4]
1991	Engineering Technology Research Centers	84 research centers (at a total cost of RMB 1.5 billion) were established in universities and GRIs to provide technical R&D and assistance to industry.[5]
1992	Climb Plan	Designed to encourage basic research efforts.[6]

Notes:
1. Annalee Saxenian and Xiaohang Quon (2005), 'Government and Guanxi: The Chinese Software Industry in Transition', *The Software Industry in Emerging Markets*, p.78.
2. *Id.*
3. Shulin Gu (1999), *China's Industrial Technology Market: Reform and Organizational Change*, p.30.
4. M.J. Greeven (2004), *The Evolution of High-Technology in China after 1978: Towards Technological Entrepreneurship*, ERTM Report Series: Research Management, 9, available at http://papers.ssrn.com/sol3/Papers.cfm?abstract_id=636798.
5. Carl J. Dahlman and Jean-Eric Aubert (2001), *China and the Knowledge Economy – Seizing the 21st Century*, p.120.
6. Greeven, *supra* note 4, at 9.

productively. In March 1985, China's State Council issued the Decision on Reform of the S&T Management System (the Decision) to create a 'technology market' to help improve the 'effectiveness' of GRIs (see Box 2.1, which provides an English version of Premier Zhao Ziyang's March 1985 speech to the National Working Conference of Science and Technology that explained the need for a technology market).

Recognizing the inherently dynamic nature of technology – where supply and demand are in a constant state of flux – the Decision sought to create a mechanism that could cope with that dynamism and effectively integrate China's GRIs with its business sector.[50] The Decision sought to create a dynamic and adaptable relationship between GRIs, the business sector and the government through the use of 'markets.' The Decision involved a number of fundamental changes to China's innovation system, including:

- *Competitive Funding*: it introduced a competitive process for the allocation of some public R&D funding (e.g, the National Natural Sciences Foundation (1986) and Key Science and Technology Projects of China's 7th Five-Year Plan (1986–1990)), which created a 'quasi-market for government funds.'[51]
- *Incentivized Responsiveness to the Needs of the Business Sector*: it reduced government funding 'to put pressure on [GRIs] so that they would have to turn to real demands,' coupled with incentive programs for both institutions and individual researchers that allowed them to 'retain a proportion of their earnings' from commercialized technology.[52]
- *Allowed Researchers to Adapt to Constantly Changing Opportunities*: it increased the decision-making autonomy of GRIs.[53] GRIs, for example, were given greater autonomy for choosing projects and hiring personnel.[54]
- *Attempted to Create a Technology Market to Disseminate Technology Developed by GRIs*: it enacted regulations and established supporting

BOX 2.1   PREMIER ZHAO ZIYANG'S MARCH
1985 SPEECH TO THE NATIONAL
WORKING CONFERENCE OF SCIENCE
AND TECHNOLOGY – EXPLAINING THE
NEED FOR A TECHNOLOGY MARKET

The current science and technology institution in our country has evolved over the years under special historical situations. The advantages embodied in this system manifested themselves in concerted efforts to tackle major scientific and technological projects, with great success. However, there is growing evidence to show that the system can no longer accommodate the situation in the four modernizations program, which depends heavily on scientific and technological progress. One of the glaring drawbacks of this system is the disconnection of science and technology from production . . .

By their very nature, there is an organic linkage between scientific research and production. For this linkage a horizontal, regular, many-leveled and many-sided channel should be provided. The management system as practiced until now has clogged this direct linkage, so that research institutes were only responsible to the leading departments above, in a vertical relationship, with no channel for interaction with the society as a whole or for providing consultancy services to production units. This is the root cause of the inability of our scientific research to meet our production needs over the years. . .

If you want scientific research to serve production needs, you must acknowledge the value created by mental labor and allow most technology achievements to become tradeable. If you want the scientific personnel of the research institutions to voluntarily and regularly go to enterprises to identify research items, you must alter the funding system in which research institutions depend entirely on appropriations from the state. To bind research institutions and production units in a common cause, you must adopt a variety of economic means linking them with ties of interests. Premier Zhao Ziyang (1985)[1]

Note:
1.   English version of speech obtained from Shulin Gu (1999), *China's Industrial Technology Market: Reform and Organizational Change*, pp.17–18.

agencies that would permit a market for technology developed by GRIs (e.g., the Law of Technological Contracts was adopted in 1987).[55]

The Decision's various initiatives took quite some time and a fair amount of experimentation to develop. Competitive funding, for example, did not instantly become the norm and replace decisions based on politics (or even corruption). In fact, even a quarter of a century later, politics probably still play too big a role in China's public R&D funding decisions.[56] The Decision's most fundamental initiative, however, which was to establish a technology market to disseminate technology developed by GRIs, did not materialize as quickly as the government had hoped. GRIs did not start selling technology to the business sector at anywhere near the levels desired by the government.

There are a number of possible explanations for the failure of the technology market to develop initially. One of the more compelling explanations was the fact that China's business enterprises at the time tended not to be very technologically sophisticated and were limited in their ability to absorb new technology.[57] GRIs, by their nature, tend to focus on incremental improvements to existing technology. In order for such incremental improvements to be useful to a technology buyer, the buyer needs to already possess expertise in the existing technology. China's businesses tended to lack such expertise, which meant they were more inclined to purchase 'packaged technology sets,' that were already prepared for use by industry.[58] Foreign technology suppliers were frequently the preferred source for such 'packaged technology sets.'[59] As a result, GRIs could not depend on the limited market as a meaningful source of revenue, which dramatically reduced their incentive to serve industry's needs. When technology buyers were available, the paucity of technology transactions likely made valuing the technology and documenting the transactions very challenging, making it much more difficult for buyers and sellers to reach agreement. A complementary explanation may be that China's institutions in the mid-1980s were not sufficiently developed to support a technology market. Faith in the intellectual property laws and confidence in contract enforceability were not particularly high in China during the 1980s. Such apprehension adds to the cost of trading in property rights and could have made many technology transactions prohibitively expensive.

4.2   Merging GRIs into Business Enterprises

In response to the disappointing technology market, China's government sought an alternative solution for transferring GRI-developed technology to the business sector. Rather than form linkages between GRIs and the business sector, the government sought to merge a number of GRIs directly into existing business enterprises. China's State Council announced this merger strategy in 1987 with its Stipulations of the State Council for Furthering the Reform of the S&T Management System. The State Council explained:

> [though the [Decision] has achieved preliminary success over the past year and more . . .] one should have been conscious that the disconnection between S&T and production has not yet been fundamentally improved. The pattern of the organization structure of the S&T system is basically untouched, the system remains closed (to the outside); the important R&D institutes are still affiliated to administrative organs rather than being bound up with the national economy; there are more qualified scientists and technicians than required in big research institutes belonging to central ministries and institutes of higher education, while there is a serious lack of S&T manpower in light industry, commercial enterprises and rural areas; the policy measures intended to intensify the links between research institutes and enterprises have been inefficient, so that a considerable number of research institutes are undertaking a kind of 'self accomplishment' [of the commercialization of their technological strengths] without devoting much effort to making outside connections.[60]

Rather than improving technology transfer from government-funded laboratories to the market place, the State Council discovered that the reforms (including the technology market solution) were causing GRIs to become even more insular. GRIs remained the primary recipient of R&D funding (they received 61% of China's R&D funding in 1987, compared to 35% for enterprises and 4% for universities),[61] but still were failing to transfer technological advances to industry. Instead, GRIs were retaining the more valuable technology and trying to commercially exploit it themselves.[62] In 1987, the government responded by launching a number of policy initiatives aimed at encouraging GRIs to merge into existing business enterprises.[63] The results of these efforts were largely disappointing, as only a few GRIs (out of the thousands that existed at the time) responded to the merger encouragements.[64] By 1988, the government had officially admitted that the merger initiative was not successful.[65]

4.3   Creating a Chinese Start-up Industry – The Torch Program

In 1988, the government tried yet another approach to increase the flow of GRI-developed technology to the commercial sector. Following the example set by the GRI's self-commercialization efforts throughout the 1980s – e.g., New Technology Company (the predecessor of Lenovo) – the government launched the Torch Program. It is worth noting that the number of GRI start-ups 'was already significant in 1988 when policy makers acted in response to them.'[66] Rather than establish new policy, the Torch Program can be more accurately described as formally recognizing (or legitimizing) a practice that had already proven to be successful. The Torch Program was launched to address two issues in particular: (i) to encourage the further creation of GRI start-ups; and (ii) to develop specific geographical areas where such start-ups could be concentrated.[67] The Torch Program included a series of policy measures to accomplish these goals, including:[68]

- GRI start-ups could be formed as new technology enterprises (NTEs), which were allowed preferential regulatory treatment.
- NTEs could hold IP rights and IP rights could be counted as equity contributions to the NTE.
- NTEs were eligible for fiscal incentives.
- NTEs received preferential financing treatment. Special financing sources (e.g., 'lending funds') were established for NTEs and banks were encouraged to finance NTEs and their commercialization efforts (see Chapter 7 for a discussion of the typical financing strategies/treatment for China's start-ups during the late 1980s and early 1990s).
- Development Zones were established to house the various NTEs. These Development Zones provided supportive environments for NTEs by providing appropriate facilities, incubator-like services and access to foreign technology and know-how. Development Zones also allowed the government to conduct a gradual, and contained, opening of China's economy to the outside world.

Unlike the earlier 'technology market' and 'GRI merger' reforms, the Torch Program proved to be quite successful at spurring technology transfer from GRIs to the market place. One commentator notes that:

[T]he Torch plan can be seen as the major breakthrough of technology in China. It is for the first time that science & technology and commercial initiatives are successfully linked.[69]

The Torch Program has left a lasting imprint on China's science and technology industry, and both GRIs and universities continue to take advantage of the program's various incentive programs.[70] The most notable impact has been the creation in the late 1980s and early 1990s of numerous high-technology parks throughout China. There are currently 54 such high-technology parks throughout China (see Table 2.2, which provides a list of China's high-technology parks). With the exception of three of China's poorest regions (i.e., Qinghai, Ningxia and Tibet), each of China's provinces now has at least one of these high-technology parks.[71]

The clustering of high-technology companies round a scientific research center frequently serves as the model for technology-based regional development. Based on the Silicon Valley model, such high-technology clustering can provide numerous advantages for high-technology companies such as increased knowledge intensity (and the resulting knowledge spillovers), a concentration of highly-skilled workers, a climate that 'rewards risk-taking and tolerates failure,' and a constrained geographic location that makes it easier for valuable relationships to form between GRIs, universities, specialized intermediaries (e.g., venture capital firms, angel investors and specialized attorneys) and high-technology firms.[72] While China's high-technology parks tend to suffer from a number of operational challenges that limit their success (and none of them operates on a level anywhere near that of Silicon Valley),[73] these clustered Chinese high-technology parks do provide China with valuable S&T infrastructure assets that have the potential for much greater exploitation in the future. Another critical benefit from the Torch Program was that it permitted the importation of valuable foreign technology and know-how into China at a time when China was much more insular to the outside world.

5.   MARKET REFORMS TAKE HOLD (EARLY 1990S)

By the end of 1992, China had established the basic framework for a modern innovation system. The linear, functional-specialization approach to innovation was being replaced with a much more dynamic, market-based system that could adapt to the constantly changing opportunities and challenges of today's knowledge-based world. GRIs understood the need to develop productive linkages with the business sector (even if they were having trouble doing so). The business sector understood that technological advancements were necessary for economic success (even if they were having trouble finding workable models for obtaining that technology). Universities in particular (and the education system in general) were playing much more important roles in the innovation system. To

Table 2.2   China's development zones (high-technology parks)

Region	Municipality/ Province	Name of High- Technology Park	Year Established	Website
Eastern Region	Beijing	Zhongguancun	1988	http://www.zgc.gov.cn
	Fujian	Fuzhou	1991	http://www.fzhitech.com
		Xiamen	1991	http://www.xmtorch.gov.cn
	Guangdong	Guangzhou	1991	http://www.getdd.gov.cn
		Shenzhen	1991	http://www.ship.gov.cn
		Zhongshan	1991	http://www.zstorch.gov.cn
		Foshan	1993	http://www.fs-hitech.gov.cn
		Huizhou Zhongkai	1992	http://www.hzzk.cn
		Zhuhai	1993	http://www.zhuhai-hitech.com
	Guangxi	Guilin	1991	http://www.eguilin.org
		Nanning	1993	http://www.nnhitech.gov.cn
	Hainan	Haikou	1991	http://www.haikou.gov.cn
	Hebei	Shijiazhuang	1991	http://www.shidz.com
		Baoding	1993	http://www.bd-ctp.net.cn
	Liaoning	Dalian	1991	http://www.ddport.com
		Shenyang	1991	http://www.hunnan.gov.cn
		Anshan	1993	http://www.asht-zone.gov.cn
	Jiangsu	Nanjing	1991	http://www.njnhz.com.cn
		Changzhou	1993	http://www.czxd.gov.cn
		Suzhou	1993	http://www.snd.gov.cn
		Wuxi	1993	http://www.wnd.gov.cn
	Shandong	Jinan	1991	http://www.jctp.gov.cn
		Weihai	1991	http://www.whtdz.com.cn
		Qingdao	1993	http://www.qdhtp.com
		Weifang	1993	http://www.wfgx.gov.cn
		Zibo	1993	http://www.china-zibo.com
	Shanghai	Shanghai Zhangjiang	1991	http://www.sh-hitech.gov.cn
	Tianjin	Tianjin Binhai	1991	http://www.thip.gov.cn
	Zhejiang	Hangzhou	1991	http://www.hhtz.gov.cn
		Ningbo	1999	http://www.nbhtp.gov.cn
Central Region	Anhui	Hefei	1991	http://www.hefei-stip.com.cn
	Heilongjiang	Harbin	1991	http://www.kaifaqu.com.cn
		Daqing	1993	http://www.dhp.gov.cn
	Henan	Zhengzhou	1991	http://www.zzgx.gov.cn
		Luoyang	1993	http://www.lhdz.gov.cn
	Hubei	Wuhan Donghu	1993	http://www.wehdz.gov.cn
		Xiangfan	1992	http://www.xfgx.com/
	Hunan	Changsha	1991	http://www.cshtz.gov.cn
		Zhuzhou	1993	http://www.zzhitech.com
	Jiangxi	Nanchang	1993	http://www.nchdz.com
	Jilin	Changchun	1991	http://www.chida.gov.cn
		Jilin	1993	http://www.jlhitech.com
	Neimenggu	Baotou	1993	http://www.rev.cn
	Shanxi	Taiyuan	1993	http://www.tyctp.com.cn

Table 2.2   (continued)

Region	Municipality/ Province	Name of High-Technology Park	Year Established	Website
Western Region	Chongqing	Chongqing	1991	http://www.hnzcq.gov.cn
	Gansu	Lanzhou	1993	http://www.lzhtp.gov.cn
	Guizhou	Guiyang	1993	http://www.gyhtz.gov.cn
	Shaanxi	Xi'an	1991	http://xaportal.xdz.com.cn
		Baoji	1993	http://www.bj-hightech.com
		Yangling	1997	http://www.ylagri.gov.cn
	Sichuan	Chengdu	1991	http://www.cdht.gov.cn
		Mianyang	1993	http://www.myship.gov.cn
	Xinjiang	Urumqi	1993	http://www.uhdz.gov.cn
	Yunnan	Kunming	1993	http://www.kmhnz.gov.cn

Sources: In addition to the individual websites for each high-technology park, Cong Cao (2004), 'Zhongguancun and China's High-Tech Parks in Transition', 44 *Asian Survey* 647, 648.

top things off, policymakers clearly understood the importance of S&T to China's economic success and were being thoughtful in their attempts to improve the innovation system. This improved framework was not just about theory – it was also generating positive, tangible results:

- China's technology creation capacities improved in its GRIs and began to develop at its research universities. See Boxes 2.2 and 2.3 for a description of two of China's biggest spin-off success stories from the 1980s. Box 2.2 provides a description of Lenovo's spin-off from the CAS and Box 2.3 a description of the Founder Group's spin-off from Peking University.
- High-technology industries and high-technology clusters began to form.
- Linkages began to develop between China's R&D efforts (including GRIs and research universities) and its commercial initiatives.
- China increased its inventory of knowledge capital – by increasing the education level of its population, by indigenous technology development and by importation of foreign technology into China.

While substantial progress was made, with a number of high profile success stories, things were by no means perfect. The overall progress towards a vibrant, well-functioning innovation system tended to be 'disjointed and uneven'[74] and many of China's deep-rooted problems from the central-planning era persisted.

Funding for R&D and education remained low by international standards, policy continued to reinforce the concentration of applied research in public research institutes rather than industry, and the weight of policy – hence the flow of resources – remained biased toward state-owned enterprises rather than the potentially more creative and innovative non-governmental technology enterprises.[75]

BOX 2.2   LENOVO – FROM US $25 000 CAS SPIN-OFF TO GLOBAL LEADER IN PERSONAL COMPUTERS

In 1984, Liu Chuanzhi and 10 colleagues from the Institute of Computer Technology of the CAS founded New Technology Developer, Inc. (the predecessor of the Legend Computer Group Corporation) with an initial cash outlay of RMB 200 000 (US $25 000) and technology developed at the CAS. Legend's initial focus was to sell, and provide after-sales service for imported personal computers and other computer products. By 1990, Legend had launched its first Legend PC in the market and was transforming its primary function to being a producer and seller of Legend-branded computer products, rather than acting as an agent for imported computer products. In 1992, Legend pioneered the home PC concept in China and became China's market-share leader in computers in 1996. Legend continued to grow as a computer-company leader in China and abroad throughout the 1990s and early 2000. In 1999, for example, Legend became the top PC vendor in the Asia-Pacific region and was No. 1 in China's national Top 100 Electronic Enterprises ranking.

Legend changed its name to 'Lenovo' in 2003 in anticipation of a major push into overseas markets, and in 2004 it acquired IBM's global PC (desktop and notebook computer) business. Today, Lenovo is a leading company in the global PC market with a full array of products and services that generated 2008 revenues of more than $16 billion. Lenovo has also developed substantial R&D capabilities and has major research centers in Beijing, Shanghai and Shenzhen, China; Yamato, Japan; and Raleigh, North Carolina.

Sources: Lenovo Website – Company History, available at www.pc.ibm.com/ca/about_lenovo/companyhistory.html. Gu, *supra* note 40, at 76–77.

BOX 2.3   FOUNDER GROUP – $50 000 UNIVERSITY SPIN-OFF THAT EVOLVED INTO AN INTERNATIONAL TECHNOLOGY CONGLOMERATE

Peking University New Technology Company (the predecessor to the Founder Group) was formed in 1986 to commercialize a patented laser typesetting system. The laser typesetting system's development grew out of a 1974 government project (the August 1974 Project) to develop a Chinese character information processing system. Peking University contributed $50 000 and the technology to Founder Group, which has gone on to become China's market leader in printing technology. Founder Group has roughly 85% market share of China's domestic printer market and exports products to more than 30 countries.

In addition to its printer prowess, Founder Group has evolved into one of China's largest technology companies, and generated US$ 5.6 billion in revenues in 2007. Founder Group serves as the holding company for five listed companies (on the Shanghai, Shenzhen, Malaysia and Hong Kong stock exchanges) as well as numerous other subsidiaries and joint ventures that provide a wide variety of technology-based products and services. Founder Group has an extensive IP portfolio (e.g., Founder Group owns, or jointly owns with Peking University, 128 Chinese patents in print and information technology) and has developed substantial R&D capacities. Founder Group also acts as a quasi-incubator for Peking University.

Source: Founder Website – About Founder, available at www.founder.com/en/About_Founder/Overview.html. Hua Guo (2007), 'IP Management at Chinese Universities', in Anatole Krattiger et al. (eds), *Intellectual Property Management in Health and Agricultural Innovation – A Handbook of Best Practices*, 1673, 1680–1681.

The period through the early 1990s did not result in a true market-oriented, merit-based innovation system. The environment was improving, but China's innovation institutions were not yet adequate and the technology intermediaries necessary to link the various actors within the system had not developed sufficiently. Incentives remained weak and the self-forming linkages that characterize a vibrant, market-based innovation system were not present. Without strong institutions and intermediaries,

China's central government remained the dominant innovation system force – even as it was trying to cede control over the innovation system to market forces. Central planning no longer controlled the innovation system, as the central government transferred much of the decision-making control to the various innovation system actors. Because these various actors were themselves frequently government-controlled actors, however, the primary decision makers in China's innovation system through much of the 1990s continued to be government-controlled entities – namely, GRIs and large, state-owned enterprises that had trouble abandoning their planned-economy practices. Privately-owned enterprises,[76] which tend to be the dominant actors in the market-based innovation systems of the developed world, were not the primary decision makers. Moreover, technology importation remained highly regulated, which restricted the flow of internationally available technology into China and slowed the growth of China's overall inventory of technological know-how.

While China did not have a market-based innovation system by the early 1990s, the foundations for such a system were being built. Significant amounts of money were being invested in R&D to increase the capacity of the country's researchers, and those funds were increasingly being spent in a merit-based fashion. China's GRI-centric innovation system was slowly being restructured into a modern, multi-faceted innovation system built on the diverse R&D strengths of its business and university sectors in addition to its GRI sector. Chinese universities were beginning to develop as meaningful technology actors, and so too were a host of Chinese start-up companies. At the same time, an institutional framework that could meaningfully support market-based interactions between China's increasingly diverse set of technology actors was also developing. China was in the process of developing a modern legal system that could support such market-based interactions. During the 1980s, for example, China created a new system of contract law, developed a system for creating and operating various types of business organizations and began to adopt modern IP laws. By the mid-1990s, with the foundations for a market-based innovation system solidly established, China's truly meaningful transformation to a market-based innovation system was able to begin.

3. A snapshot of China's current innovation system

[China plans] to join the ranks of innovative countries [by 2020], thus paving the way for China to become a world leader in science and technology by the middle of the 21st century. – China's National Medium- and Long-Term S&T Development Plan (2006–2020)[1]

Ideally, countries that employ a Bayh–Dole strategy to encourage universities to develop and commercialize technology will possess a strong university sector that has the capacity to generate useful, commercial technology and a vibrant business sector that is capable of absorbing that technology. This chapter examines the current roles and capacities of China's university and business sectors, as well as their ability to link their activities and contribute to a Bayh–Dole strategy. This chapter will also briefly examine China's restructuring of its GRIs in 1999 and 2000, as that was one of China's more significant innovation system reforms and helps to place in context how Chinese universities and companies function in China's innovation system.

1. CHANGING ROLES FOR CHINA'S FUNDAMENTAL TECHNOLOGY ACTORS

China's innovation system continued to evolve throughout the 1990s and 2000s and has experienced some of the more profound changes in China during that period. Some of the more significant reforms include:[2]

- *GRI Sector*: the CAS was restructured to focus its efforts on early stage R&D (e.g., basic and applied research) and to upgrade the quality of its personnel. At the same time, government support for thousands of other GRIs was eliminated and those GRIs were privatized.
- *University Sector*: China took the novel approach of focusing its education reform on higher education, rather than on primary or secondary education – which is the more traditional strategy for

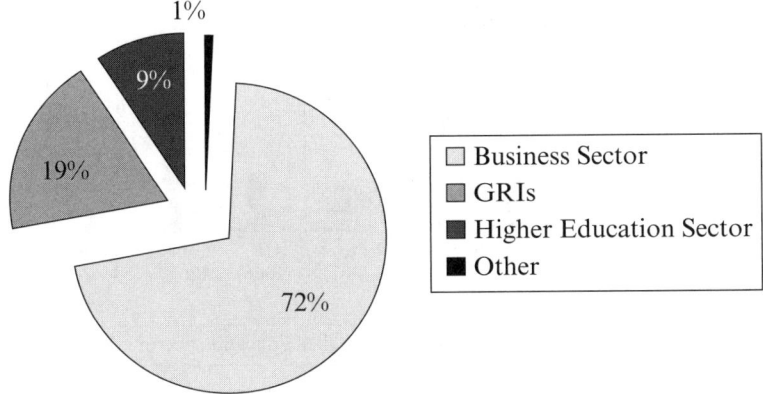

Source: MOST (2008), *China S&T Statistics Databook*, available at www.sts.org.cn/sjkl/ kjtjdt/data2008/cstsm08.htm. MOST (1998–2007), *China S&T Statistics Databooks* are available at www.most.gov.cn/eng/statistics/2007/index.htm.

Figure 3.1   China's distribution of R&D expenditures by sector – 2007

developing countries.[3] China expanded its higher education sector to dramatically increase its number of undergraduate and gradu-ate students. China also sought to create a select number of elite research universities by focusing its financial support toward them.

- *Business Sector*: China recently announced that the business sector will be the driving force for its R&D efforts, and available statis-tics generally show that the business sector has taken on that role. China's business sector is the largest R&D performer based on both R&D inputs (e.g., R&D expenditures) and outputs (e.g., patent applications, patents granted, and license revenue).[4] In spite of the ever increasing role of the business sector in China's innovation system, many continue to question the sufficiency of the business sector's innovation capacity,[5] which retards its ability to absorb university-developed technology.

R&D in China has evolved from being the near monopolistic domain of GRIs to being very much of a shared endeavor by GRIs, research universi-ties and the business sector (see Figure 3.1, which sets forth China's distri-bution of R&D expenditures by sector). Overall, China's division of R&D expenditures between the business sector, the higher education sector and GRIs is now quite comparable to the world's leading S&T countries (see Figure 3.2).

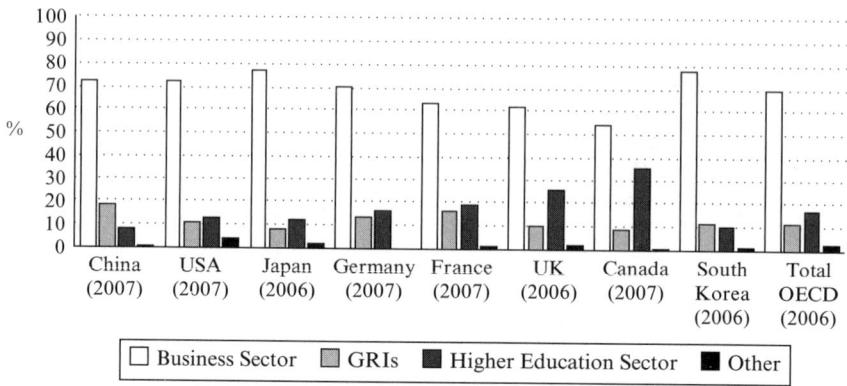

Sources: MOST (2008), *China S&T Statistics Databook*; and OECD (2008), *Main Science and Technology Indicators*, Vol. 2.

Figure 3.2   International comparison of the distribution of R&D expenditures

2.   RESTRUCTURING OF GRIS

China fundamentally restructured its GRI sector during the late 1990s and early 2000s. GRIs have gone from being the centerpiece of China's innovation system to playing a complementary role that is similar to the role of GRIs in many developed countries. China's GRIs are populated by high-quality research personnel who focus primarily on basic and applied research, and on research with a 'public-goods quality' like agriculture and defense.[6] Stated another way, GRIs focus on research areas where market forces are most likely to motivate the private sector to under invest.

The CAS, which is the undisputed crown jewel of China's GRI sector, has undergone numerous reorganizations to refocus its work on priority areas and upgrade the quality of its personnel.[7] Non-CAS GRIs were subjected to even stronger reforms. In 1999, 242 GRIs attached to ten bureaus under the State Economic and Trade Commission were transformed into private enterprises.[8] This 1999 privatization effort was followed in 2000 by an announcement that China would require its 4000 regional GRIs to financially support themselves. Some of the privatized GRIs merged with existing industrial enterprises, while some sought to operate as stand-alone entities.[9] While it is unclear if all of the 4242 GRIs were actually privatized, it is clear that a substantial number of them were converted into industrial enterprises. The industrial conversion sought to define more precisely the role of GRIs in China's economy and to better focus government R&D spending. GRIs with lesser qualifications would no longer be subsidized

Table 3.1   Decrease in number of GRIs and GRI personnel following industrial conversion

	1991	1998	2003
Number of institutes	5867	5778	4169
Number of S&T personnel	800 000	590 000	410 000
Number of employees	1 070 000	940 000	570 000

Source: Martin Schaaper (2009), 'Measuring China's Innovation System: National Specificities and International Comparisons', *OECD Science, Technology and Industry Working Papers*17, available at www.oecd.org.dataoecd/15/55/42003188.pdf. (citing MOST, *The Yellow Book on China Science and Technology Vol. 7*, 2004, Figure 3–4 (2005)).

by the government, while stronger GRIs with capacity in early-stage and 'public-goods quality' research could receive greater financial support.[10] So, while the number of GRIs and GRI S&T personnel has decreased (see Table 3.1), the quality appears to have improved.[11]

3.   UNIVERSITY SECTOR

China's universities play a central role in China's innovation system. First, China's universities serve the traditional role of educating the most technologically advanced segment of the population so they can play more significant roles in the innovation system. A country cannot develop a successful, technology-based economy without a critical mass of highly-skilled workers (e.g., scientists, engineers and technical workers) that are specifically trained to operate in science and technology fields. China's higher education reforms throughout the 1990s dramatically increased the number of students able to obtain an undergraduate or graduate-level university education.

Universities also play a much more direct role in the innovation process by serving as a valuable source of technological innovations. China's higher education reforms have also sought to develop a select number of elite research universities that are capable of generating valuable technological innovations that can contribute to China's economic development.

3.1   Overview of China's Higher Education System

China possesses an extensive higher education system that consists of PhD-granting universities and GRIs, four-year universities and colleges

Table 3.2   Number of Chinese higher education institutions in 2004

	No. of Institutions
Institutions offering PhDs	**769**
Universities	454
GRIs	315
Regular Tertiary Education Institutions	**1731**
Four-year universities and colleges offering bachelor degrees	684
Three-year colleges offering associate degrees	175
Vocational technical colleges	872

Sources: OECD (2007), Thematic Review of Tertiary Education – Background Report for the P.R. of China 14 and Uwe Brandenburg and Jiani Zhu (2007), 'Higher Education in China in Light of Massification and Demographic Change', *Arbeitspapier NR.* 97, at 18.

that offer bachelor degrees, two- and three-year colleges (frequently referred to as 'short-cycle' colleges) that offer associate degrees, and vocational technical colleges.[12] See Table 3.2, which provides statistics on the number of Chinese higher education institutions in 2004.

In many statistical classifications, China's adult education schools are also listed as being part of the tertiary education system. China's adult education system includes a range of programs that typically focus on such matters as worker and peasant elementary schools, adult literacy classes and worker training courses.[13] We have not included China's adult education schools in Table 3.2.

Enrollment in Chinese universities began to rapidly increase in 1999, after years of relatively stable enrollment and graduation numbers.[14] China's undergraduate and postgraduate enrollment numbers have increased roughly 30% per year since 1999, resulting in a nearly six-fold increase in the number of Chinese students at the tertiary level (see Figures 3.3 and 3.4). This massive increase in the pool of undergraduate and postgraduate students has generated a corresponding increase in graduating students beginning in 2002 (see Figures 3.3 and 3.4). Four and one-half million students received undergraduate degrees in China in 2007 and 256 000 received postgraduate degrees in 2006. In 1998, by comparison, 830 000 undergraduate degrees and 47 000 postgraduate degrees were issued.

Unlike most OECD countries, science and engineering majors are 'held in esteem' in China, rather than liberal arts studies.[15] Roughly one-quarter of graduates from OECD universities obtain a degree in science or engineering.[16] By comparison, that number was 41% in China in 2007.[17] When

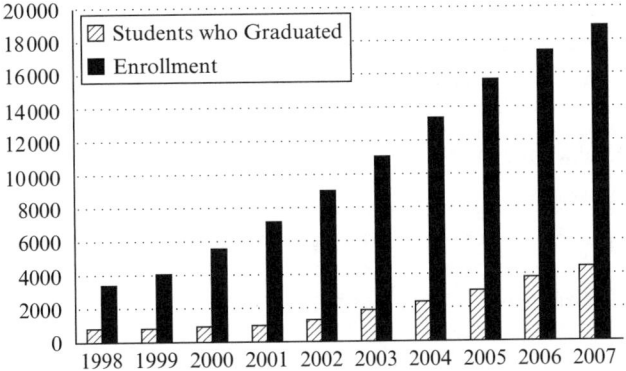

Source: MOST (1999–2008), *China S&T Statistics Databooks.*

Figure 3.3   Enrollment and graduation statistics for undergraduates at China's regular institutions of higher education (thousands of students)

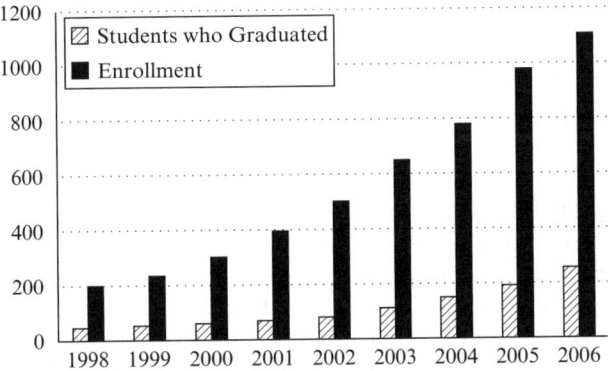

Source: Ministry of Education of the PRC.

Figure 3.4   Enrollment and graduation statistics for postgraduate students at China's regular institutions of higher education + GRIs (thousands of students)

coupled with the substantial increase in tertiary enrollment and China's immense population, China now annually produces in excess of 1.5 million science and engineering graduates from its undergraduate institutions – with the bulk on the engineering side.[18] Such statistics, however, come

Table 3.3   Postgraduate science and engineering students at China's regular institutions of higher education + GRIs

	1998	2000	2002	2004	2006
Science & Engineering PhDs					
Enrollment:					
Science	9253	12556	18692	28769	35884
Engineering	*21553*	*31284*	*46496*	*69315*	*86833*
Total	30806	43840	65188	98084	122717
PhD Degrees Granted:					
Science	2284	2408	2808	4518	7241
Engineering	*3427*	*4611*	*5252*	*8054*	*12130*
Total	5711	7019	8060	12572	19371
Science & Engineering Masters					
Enrollment:					
Science	20353	29177	45248	73612	98845
Engineering	*62867*	*98114*	*150782*	*248748*	*325440*
Total	83220	127291	196030	322360	424285
Masters Degrees Granted:					
Science	5189	5669	7058	13022	21896
Engineering	*17254*	*19752*	*24826*	*48020*	*82386*
Total	22443	25421	31884	61042	104282

Source: Ministry of Education of the PRC.

with a big qualification. Graduates from three-year technical colleges make up close to half of these science and engineering graduates.[19] These technical colleges are roughly equivalent to U.S. community colleges, and it is a stretch to classify their graduates as scientists or engineers.[20] They are probably more appropriately classified as skilled technicians. Nevertheless, even if China is graduating *only* 750000+ science and engineering undergraduates each year, that number is still 50% greater than the United States' annual output of roughly 500000 science and engineering undergraduates.[21]

China's progress has been no less impressive at the postgraduate level (see Table 3.3, which provides enrollment and graduation statistics for postgraduate science and engineering students in China). From 2000 to 2006, China almost tripled its output of science and engineering PhDs and more than quadrupled its output of science and engineering masters' degrees. Some have predicted that China will have more science and engineering PhDs than the United States by 2010.[22] In 2006, China granted more than 19000 science and engineering PhDs, which compares favorably to the United States' issuance of just under 16000 PhDs in physical·sciences and engineering in 2007.[23]

3.2 Developing Elite Universities

In addition to the amazing growth in numbers, China has undertaken a number of significant higher education reforms that tend to receive less attention, but could be equally important in considering China's technology future. First, China has sought to increase its emphasis on the 'quality' of university education.[24] To help ensure professor quality, China has increasingly turned to benchmarking and other objective quality indicators – such as publications in internationally recognized journals, citation rates, and international collaborations – to guide funding decisions as well as faculty hiring, advancement, and retention decisions.[25] Chinese university professors tend to be subject to regular reviews that go beyond the tenure-review process employed in many academic settings.[26] In fact, '[i]t is not uncommon for an annual target of three international publications to be set for faculty members, with termination of employment to occur on non fulfillment.'[27]

At the same time, China has sought to promote so-called 'elite' universities that can compete on an international level in the quality of their programs and research. Two of the more important reform efforts to establish such elite universities are Project 211, which was launched in 1995, and Project 985, which was launched in 1999. Project 211 sought to build roughly 100 top level universities in the twenty-first century and provided them with additional funding for such things as developing key disciplines, building support facilities, and hiring faculty.[28] A list of the 107 universities and colleges that are 'Project 211' institutions (as of 2007) is included in the notes to this chapter.[29] While representing only 6% of China's 1700 regular tertiary education institutions, Project 211 institutions educate roughly 80% of China's doctoral students, 67% of its graduate students, 50% of its foreign students and 33% of its undergraduates.[30] Project 211 institutions also possess 96% of China's key laboratories and receive 70% of its scientific research funding.[31] There appear to be conflicting figures as to how much extra funding has been allocated to Project 211 institutions. A March 2008 statement by Zhao Lu, director of the Education, Science and Culture Department under the Ministry of Finance, however, announced the following investment of special funds by the government in Project 211 institutions:[32]

- Phase one (1996–2000): RMB 2.8 billion
- Phase two (2001–2005): RMB 6 billion
- Current phase (2006–2010): RMB 10 billion

Table 3.4   China's 39 Project 985 institutions

The 9 Original Institutions

Peking University	Harbin Inst. of	Univ. of Science and
Tsinghua University	Technology	Tech. of China
Fudan University	Nanjing University	Xi'an Jiaotang University
	Shanghai Jiaotang	Zhejiang University
	University	

The 30 Institutions Added to the Program During Phase Two

Beihang University	Hunan University	Shandong University
Beijing Institute of	Jilin University	Sichuan University
Technology	Lanzhou University	South China Univ. of
Beijing Normal University	Nankai University	Technology
Central South University	National Univ. of	Southeast University
Central University for	Defense Tech.	Sun Yat-Sen University
Nationalities	Northeastern University	Tianjin University
China Agricultural	Northwestern	Tongji University
University	Polytechnical Univ.	University of Electronics
Chongqing University	Northwest Sci-Tech	Science & Tech.
Dalian University of	Univ. of Agriculture	Wuhan University
Technology	& Forestry	Xiamen University
East China Normal	Ocean University of	
University	China	
Huazhong Univ. of	Renmin University of	
Science & Tech.	China	

Source: The Central People's Government of the PRC (in Chinese), available at http://www.gov.cn/fwxx/2009gk/content_1314252.htm.

Project 985, also referred to as 'The Project for Founding World-class Universities,'[33] is even more ambitious than Project 211 at developing truly 'elite' universities. As originally formulated, Project 985 was to cover only two universities: Peking University and Tsinghua University. The original two universities ended up being expanded to nine, however, as Fudan University, Shanghai Jiaotang University, Nanjing University, Zhejiang University, Xi'an Jiaotang University, University of Science and Technology of China, and Harbin Institute of Technology were also included as original Project 985 institutions.[34] One source estimates that Peking University and Tsinghua University received $218 million in extra funding from 1999 to 2001 as a result of Project 985.[35] During the project's second phase (2003–2007), the list of Project 985 institutions has been expanded to 39 (see Table 3.4 for a complete list of Project 985 institutions).

It appears that a third and fourth phase are intended for Project 985, as well as additional phases after that.[36] Unfortunately, reliable public information on the additional funding that flows to these Project 985 institutions is not available.

China's elite universities are beginning to receive the desired international acclaim sought by Projects 985 and 211. In 2009, for example, six Chinese universities (excluding Hong Kong universities) were ranked in the Top 200 universities in the world by the QS World University Rankings, with Peking University and Tsinghua University being ranked in the Top 60.[37]

3.3 Scholarly Output

A useful indicator of the R&D capacity of a country's universities is the success of its faculty in international scientific publications. Publishing technological advances through academic papers remains a common and effective strategy for R&D output. While knowledge transfer through publication may be less direct (and its effect more difficult to measure) than a commercial output strategy – such as patenting the technology – publications can lead to substantial improvements in a country's intellectual capital and, in many cases, remains the most effective means for diffusing new knowledge. Advances in basic research, for example, that are still far from ready for commercial application are typically diffused most effectively through a publication strategy.

Historically, China's scientists have not been active publishers in internationally recognized scientific journals. In 1990, China produced 6285 science and engineering articles, or 1.2% of the worldwide science and engineering articles, compared to 191 559 for the United States or 149 032 for the European Union.[38] In the mid-1990s, however, China's scholarly production began to increase significantly with China's annual output of science and engineering articles increasing nearly sevenfold from 1990 to 2005. See Table 3.5 for select statistics on science and engineering article output by country/region, including China, in 1995 and 2005.

Over the last decade, China's scientific institutions – research universities and GRIs – have made a number of structural changes that help explain the improved scholarly production. First, the incentive structure has changed. PhD candidates 'are now expected to publish at least one article in a journal listed in Thomson's *Science Citation Index*.'[39] With the substantial growth in China's university system and PhD candidates, this publication requirement has had a major impact on China's publication statistics. In addition, publication records are now being used more to determine funding for more experienced academics,[40] and monetary prizes (and presumably increased compensation) are now available for

*Table 3.5   Science and engineering article output by country/region:
1995–2005*

Country/Region	1995		2005		Avg. Annual Increase in No. of Articles
	No.	Share	No.	Share	
All countries	564 645	100.0%	709 541	100.0%	2.3%
United States	193 337	34.2%	205 320	28.9%	0.6%
European Union	195 897	34.7%	234 868	33.1%	1.8%
Japan	47 068	8.3%	55 471	7.8%	1.7%
South Korea	3 803	0.7%	16 396	2.3%	15.7%
India	9370	1.7%	14 608	2.1%	4.5%
China	9061	1.6%	41 596	5.9%	16.5%

Source: National Science Board (2008), Science and Engineering Indicators 2008, 5–38,
available at www.nsf.gov/statistics/seind08/.

publishing in highly-ranked journals.[41] In addition to improved incentive
structures, other factors that are likely to have encouraged this significant
increase in scholarly production include:

- A general growth in China's tertiary education, which has included a
  substantial increase in the number of higher education sector R&D
  personnel.[42]
- An increase in the number of Chinese scholars who returned from
  abroad.[43] Presumably, many of these returning scholars will have
  operated in a culture that valued academic scholarship while abroad
  and will bring that culture back with them to China.
- China's science and engineering scholars may be benefiting from the
  'demonstration effect' as more individuals develop into successful
  scholars and can serve as role models for others.
- One must even consider such mundane factors as an increase in
  English proficiency throughout China generally, which has substantially increased the ability of Chinese researchers to meaningfully
  participate in international academic circles and exchanges.

Unfortunately, the quantity of scholarly output appears to have out-
paced the quality. Paul Evans, Vice President of Elsevier (a leading pub-
lisher of scientific journals) in China, explains:

The pressure on PhD students has led to a flood of poor quality papers, which
can distract [*sic*] from the top quality work that's being done. It's a headache

Table 3.6   *Relative citation of science and engineering articles*

Country/ Region	1992		1997		2003	
	All countries/ economies	Literature from own economy excluded	All countries/ economies	Literature from own economy excluded	All countries/ economies	Literature from own economy excluded
United States	1.369	1.000	1.353	1.016	1.363	1.026
European Union	0.964	0.660	0.977	0.689	0.992	0.737
Japan	0.870	0.570	0.805	0.539	0.832	0.575
South Korea	0.410	0.290	0.480	0.323	0.620	0.439
China	0.332	0.231	0.387	0.247	0.519	0.304
India	0.280	0.150	0.317	0.196	0.439	0.284

Source: National Science Foundation, Division of Science Resource Statistics (2007), *Asia's Rising Science and Technology Strength, Comparative Indicators for Asia, the European Union and the United States*, available at www.nsf.gov/statistics/nsf07319.

for our journal editors . . . and I think it's been bad for the overall reputation of Chinese science. Typically we reject around 50 per cent of the papers we receive from the US, and around 80 per cent from China.[44]

A large portion of the world's science and engineering articles are never cited in another article,[45] which calls into question the value of the knowledge that is being transferred through such non-cited literature. More heavily cited articles tend to demonstrate higher quality R&D. For Chinese scientists, their citation rates have been relatively low by international standards. For example, '[f]rom 1993 to 2003, there were no Chinese in the top 20 most cited international scientists and only two in the top 100'.[46] Table 3.6 provides the relative citation index of science and engineering articles originated in certain select countries.[47] The relative citation index measures the 'per article' citation frequency of a county's scientific literature.[48]

China's citation index remains low (in particular when compared to the United States, the European Union or Japan), but it has shown consistent improvement. When the increasing citation index is combined with the substantial increase in the overall number of Chinese-authored articles, it is clear that China's academics have shown marked improvement. Moreover, China's science and engineering academics are becoming much more connected with the international science and engineering community (see Table 3.7), which should further boost the quality of their R&D efforts.[49]

Table 3.7   International collaboration: top international partner-countries for China's scientific authors

Country/Region	1995	2000	2005
		Number of Collaborated Papers	
United States	914	2411	5995
Japan	377	1082	2411
Germany	309	694	1422
United Kingdom	227	596	1401
Canada	198	418	1175
Australia	109	382	1024
France	174	325	866
Singapore	Not available	299	799
Korea	Not available	Not available	712

Source: OECD (2007), *OECD Reviews of Innovation Policy: China – Synthesis Report* 40 (citing to R.N. Kostoff et al. (2007)).

Finally, despite questions about the overall quality of China's science and engineering articles, there are fields of excellence within China's science community that are more than capable of competing at the very highest international level. China has top institutions that do produce highly-cited, high-impact scientific articles (see Table 3.8) – in particular in such hard science areas as physics, chemistry, engineering, material science and mathematics.[50] Not surprisingly, each of the universities from Table 3.8 is a Project 985 institution.

3.4   Substantial Increase in R&D Expenditures and R&D Personnel for the Higher Education Sector

Another useful indicator of the R&D capacity of a country's universities is the amount of R&D expenditure in the sector. China's universities have benefited from substantial growth in R&D expenditures. From 1998 to 2007, R&D expenditures in the higher education sector increased almost sevenfold (see Figure 3.5), from RMB 5.7 billion in 1998 to RMB 31.5 billion in 2007.

Government funding makes up a majority of the R&D expenditures for the higher education sector – roughly 55% each year from 2004 through 2007 – and has been the biggest reason for the strong growth in such expenditures (see Table 3.9). When coupled with the increased funding to China's 'elite' universities through Project 211 and Project 985, it is likely that much of these increased R&D expenditures have been

Table 3.8   *Chinese institutions (excluding Hong Kong) that produce the most science and engineering articles: 2004–2005*

Institution	No. of Papers
Chinese Academy of Sciences	7029
Tsinghua University	1886
Zhejiang University	1477
Peking University	1391
Shanghai Jiaotang University	1204
University of Science and Technology, China	943
Nanjing University	940
Fudan University	905
Shandong University	672
Jilin University	650
Huazhong University Science and Technology	591
Harbin Institute of Technology	590
Nankai University	581
Wuhan University	562
Xi'an Jiaotang University	533

Source: Table reproduced from Ronald N. Kostoff et al. (2006), 'The Structure and Infrastructure of Chinese Science and Technology', *Report of Office of Naval Research*, 11 and 12, available at http://fas.org/irp/world/china/docs/science.pdf.

concentrated in China's top research universities. One recent report indicated that China's 'top 50 universities accounted for 66% of total R&D expenditure in natural sciences and engineering in the higher education sector.'[51]

While not the leading source of R&D expenditures for the higher education sector, it is worth noting that the business sector has become a very significant R&D funding source – accounting for more than one-third of those funds from 2004–2007. The increased funding from the business sector demonstrates its increasingly strong linkage with the university sector. Presumably, the increased funding from the business sector has resulted, at least partially, from universities increasing their technology commercialization efforts (see Chapter 6).

Coinciding with the increase in R&D expenditures has been a comparable increase in R&D personnel in the higher education sector. Over the last three years, for example, higher education sector R&D personnel increased 11.9% from 227 000 in 2005 to 254 000 in 2007.[52]

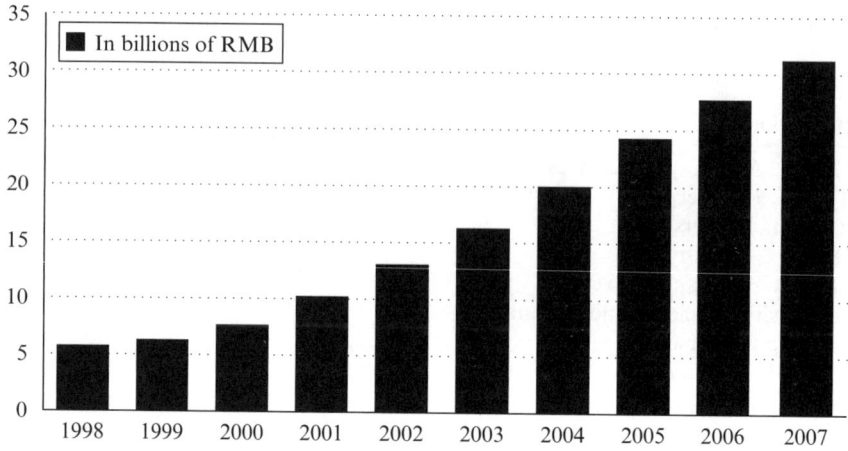

Source: MOST (1999–2008), *China S&T Statistics Databooks.*

Figure 3.5   R&D expenditures for China's higher education sector

Table 3.9   Sources of funds for higher-education R&D expenditures (in billions of RMB)

Source of funds	2004		2005		2006		2007	
	Amount	**%**	**Amount**	**%**	**Amount**	**%**	**Amount**	**%**
Government	10.9	54.2	13.3	55.0	15.2	54.9	17.8	56.5
Business	7.5	37.3	8.9	36.8	10.1	36.5	11.0	34.9
Sources from abroad	0.3	1.5	–	–	0.4	1.4	0.5	1.6
Other	1.5	7.5	0.2	0.8%	2.0	7.2	2.2	7.0

Source: MOST (2006–2008), *China S&T Statistics Databooks.*

3.5   University Growth in Basic and Applied Research

Ideally, a country's R&D efforts should include a complementary mix of basic research, applied research and experimental development (see Box 3.1 for definitions).

China's overall investment in basic and applied research remains very low by advanced S&T country standards (see Table 3.10). On a percentage basis, China's commitment to basic research is roughly one-third of the proportion committed by the United States, Japan and South Korea, and

---

BOX 3.1   DEFINITIONS FOR BASIC RESEARCH, APPLIED RESEARCH AND EXPERIMENTAL DEVELOPMENT

The OECD's *Frascati Manual*[1] provides standard definitions for the three types of research:[2]

- *Basic research* – Experimental or theoretical work under-taken primarily to acquire new knowledge of the underlying foundation of phenomena and observable facts, without any particular application or use in view.
- *Applied research* – Original investigation undertaken in order to acquire new knowledge. It is, however, directed primarily towards a specific practical aim or objective.
- *Experimental development* – Systematic work, drawing on existing knowledge gained from research and/or practical experience, which is directed to producing new materials, products or devices, to installing new processes, systems and services, or to improving substantially those already produced or installed.

Notes:
1. OECD (2002), *Frascati Manual: Proposed Standard Practice for Surveys on Research and Experimental Development.*
2. *Id.* at 30.

---

less than one-fourth of France's commitment. China does only slightly better in applied research and substantially trails each of those countries' commitment.

When looked at on an absolute basis, China has yet to show a true commitment to basic or applied research (see Table 3.11). From 2004 to 2007, for example, the R&D expenditures to basic and applied research remained largely flat, with the true growth to China's R&D expenditures coming almost entirely from increased experimental development. China's lack of commitment to more fundamental research is curious, and its consequences to China's overall innovation system are not entirely clear. The paucity of more fundamental research would appear, however, to be a negative indicator on the overall quality of China's R&D efforts.

On a more positive note, China's research universities have shown a considerable increase in their investment in more fundamental research.

Table 3.10   International comparison – R&D expenditures by type of research

	China (2007)	USA (2006)	Japan (2005)	France (2005)	S. Korea (2006)
	(as a percentage of total R&D)				
Basic research	4.7%	18.6%	12.7%	23.7%	15.2%
Applied research	13.3%	23.1%	22.2%	39.0%	19.9%
Experimental development	82.0%	58.3%	65.2%	37.3%	65.0%
	(measured in billions of US dollars)				
Basic research	$2.3	$63.9	$18.9	$11.3	$4.4
Applied research	$6.5	$79.4	$33.0	$18.5	$5.7
Experimental development	$40.0	$200.4	$96.8	$17.7	$18.6

Source: MOST (2008), *China S&T Statistics Databook.*

Table 3.11   China's R&D expenditures by type of research (in billions of RMB)

	2004		2005		2006		2007	
	Amount	**%**	**Amount**	**%**	**Amount**	**%**	**Amount**	**%**
Basic research	11.8	6.0	13.2	5.4	15.6	5.2	17.4	4.7
Applied research	40.1	20.4	43.4	17.7	50.5	16.8	49.4	13.3
Experimental development	144.7	73.6	188.7	77.0	234.2	78.0	304.2	82.0

Source: MOST (2005–2008), *China S&T Statistics Databooks.*

Different actors tend to be better suited to different research activities. Basic research, for example, will frequently generate little immediate commercial value, but may generate very high overall social returns. In such a scenario, a GRI or a government-funded research university will likely be the more logical source for such basic research, rather than the business sector. As research gets closer to product implementation, the business sector tends to be the more logical source for the R&D. While China's overall commitment to more fundamental research has been weak, China's universities have been ramping up their efforts in basic research

Table 3.12   Higher education's R&D expenditures by type of research

	2004	2005	2006	2007	% increase from 2004 to 2007
	(in billions of RMB)				
Basic research	4.8	5.7	7.1	8.7	81.3%
Applied research	10.9	12.5	13.7	16.2	48.6%
Experimental development	4.4	6.1	6.8	6.6	50.0%

Source: MOST (2005–2008), *China S&T Statistics Databooks.*

and applied research (see Table 3.12). In 2007, basic and applied research accounted for 27.5% and 51.4%, respectively, of R&D expenditures by China's higher education sector compared to 21.0% for experimental development. By comparison, those percentages in 2004 were 23.9% for basic research, 54.1% for applied research, and 22.0% for experimental development.[53]

The increased focus on more fundamental research at the university level is a positive indicator of the growing R&D strength for China's universities and their potential for generating truly novel, patentable technology.

3.6   More Direct Measures of Commercial R&D Strength

Scholarly output, increased R&D expenditures and personnel, and growth in basic and applied research provide useful, but indirect, measures for the capacity of China's universities to develop commercially viable technology. Universities' success in generating patent applications and grants, generating revenue from licensing or assigning patents, and forming successful university-based high technology start-ups provide much more direct measures. China's elite universities have shown significant improvement in each of these measures, which will be discussed in Chapter 6. Overall, there is little reason to doubt that China's elite universities have the R&D strength to be meaningful participants in a Bayh–Dole strategy.

4.   BUSINESS SECTOR

A business sector that is equipped with strong, integrated R&D capabilities is almost as important to the technology commercialization efforts of China's universities as the universities themselves. At the most basic level, the business sector needs to be technologically sophisticated enough to absorb the technology that universities seek to commercialize. Patent licensing is the most common technology commercialization strategy employed by universities in the West, but that strategy requires sophisticated recipients of the technology. Firms, or entrepreneurs, that are experienced in developing technology and have sufficient in-house R&D capacity are better positioned to understand, and capitalize on, patented technological advances that may come out of a university. For example, not all patented technology is ready-made for immediate commercialization, and it frequently requires significant R&D investment by the licensee to render the technology commercially viable. Firms without such R&D capacity are effectively eliminated as useful patent licensees.

On the positive side, China has announced that the business sector will be the driving force for its R&D efforts. Making the business sector the centerpiece of China's overall innovation system is a laudable goal for Chinese policymakers. In highly-developed countries, the business sector tends to play the dominant role in their innovation systems. In each of the G-8 countries and most OECD countries, for example, the business sector accounts for the majority of the respective country's R&D,[54] and is also the primary mechanism for putting technology to productive use by developing and selling technology-based products and services. The business sector's dominant role should not be surprising, as it provides a number of allocation advantages to an innovation system when compared to a predominantly government-driven system. The business sector is highly motivated to participate in productive R&D activities because it is the actor that is most directly able to capture the economic benefit from technological advances. It is the business sector that reduces technology to usable products and services and then sells those products and services. Of equal importance is the business sector's close proximity to the general public's needs and desires. As the actor that most directly profits (or loses money) based on the technology decisions it pursues, the business sector receives extremely valuable information on what technologies will be most useful to, and most desired by, the general public. The discipline of having to deal with customers helps the business sector to focus its resources on the most valuable technology projects.

4.1   Rapid and Self-Funded Increase in R&D Expenditures

The largest increase in R&D expenditures has been enjoyed by China's business sector, which accounted for 72.3% of China's R&D expenditures in 2007,[55] compared to 44.8% in 1998.[56] During that ten-year period, R&D expenditures in the business sector increased tenfold from just under RMB 25 billion to nearly RMB 270 billion (see Figure 3.6).

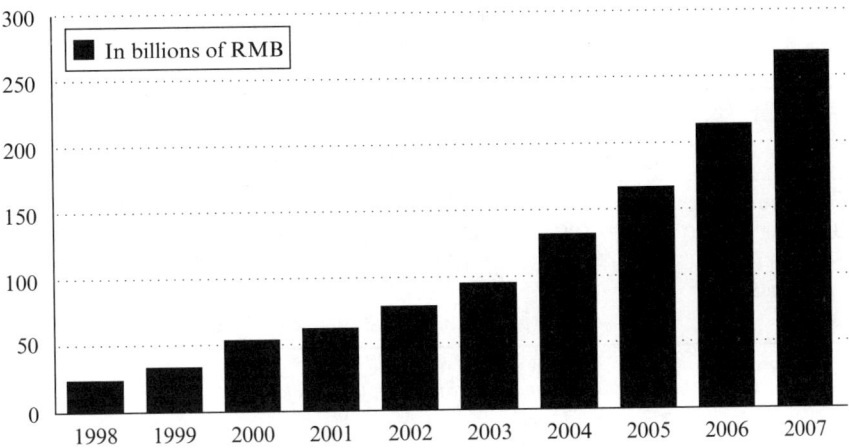

Source: MOST (1999–2008), *China S&T Statistics Databooks.*

Figure 3.6 R&D expenditures for China's business sector

Making the business sector's R&D spending increase more remarkable has been the source of funds for the increase (see Table 3.13). The vast majority of the spending increase has been self-financed by the business sector. In each year from 2004 to 2007, for example, self-funding accounted for more than 90% of the business sector's R&D funds. Looking at the issue from a slightly different angle, business sector R&D expenditures doubled between 2004 and 2007 – going from RMB 131.4 billion to RMB 268.2 billion – and self-funding accounted for 93.3% of that RMB 136.8 billion increase.

An important caveat about the self-funding statistics needs to be made. State-owned enterprises continue to dominate China's business sector, in particular the larger companies that are the most likely to have significant in-house R&D operations. It is unclear how much government subsidies to these SOEs skews the self-funding statistics.

4.2 Quality of Business Sector R&D Remains Questionable

While spending by the business sector has dramatically increased and the business sector now receives the dominant share of China's R&D resources, a closer examination of China's business sector R&D capabilities is less complimentary. Table 3.13 only shows an increase in the 'quantity' of R&D conducted by the business sector. It tells us nothing about the quality of the R&D that has resulted from the increased efforts. From

Table 3.13   Sources of funds for business sector R&D expenditures (in billions of RMB)

Source of funds	2004		2005		2006		2007	
	Amount	%	Amount	%	Amount	%	Amount	%
Government	6.3	4.8	7.7	4.6	9.7	4.5	12.9	4.8
Business	118.9	90.5	152.7	91.2	194.6	91.2	246.6	91.9
Sources from abroad	2.0	1.5	1.7	1.0	4.2	2.0	4.2	1.6
Other	4.2	3.2	5.3	3.2	5.0	2.3	4.6	1.7

Source: MOST (2006–2008), *China S&T Statistics Databooks.*

a quality standpoint, unfortunately, it appears that the R&D capability of China's business enterprises remains questionable,[57] and many continue to question whether Chinese businesses 'will develop R&D capabilities in support of novel, science-based technologies in the near future.'[58] Consider the following observations about the quality of China's business sector R&D capabilities:[59]

- In a 2007 book on high-technology research in China, Linda Jakobson noted that the 'overwhelming majority' of China's oft-cited business success stories do not owe their success to cutting-edge high technology. Top Chinese companies that have been able to list on the Nasdaq stock market, for example, have been successful based on mastering 'a competitive advantage in assembly, design, software, and/or systems engineering,' rather than technology-driven product development. The innovative strength of these companies comes from business processes, 'not creating core technology.'[60] Jakobson also noted that more than three-quarters of China's large- and medium-sized enterprises do not have R&D departments.[61]
- Less than one-third of large- and medium-sized industrial firms have their own R&D units.[62]
- Rather than focus on truly cutting-edge, transformative 'research', China's business sector has instead spent its efforts on product 'development.'[63] They have focused almost entirely on the 'D,' and not the 'R,' in R&D.
- Based on the 2009 EU Industrial R&D Investment Scoreboard, only 15 Chinese companies made the list of the top 1000 non-EU R&D investors, and only one Chinese company was listed in the top 100.[64]

- R&D intensity in most of China's high-technology industries 'is not substantially higher than in manufacturing on average.'[65]
- The business sector continues to be very reliant on purchasing foreign technology.[66]

There are a variety of explanations for the business sector's apparent tendency to take a 'lower quality' approach to R&D. Jakobson offers one of the more common explanations:

> When competition is fierce and profits are razor-thin, as is the case in many sectors in China, there is not much left over to invest. The growth of many Chinese companies has been based on cheap labor, a vast domestic market, and imported technology. Even if the technology has often been 'second-hand' – not cutting-edge technology – it has been sufficient and has reduced the need for setting up in-house R&D. Purchasing proven technology poses less risk than pursuing one's own technology. . . Chinese companies tend to focus entirely on the short-term, reflecting a degree of uncertainty that the Chinese feel about their rapidly transforming society and the possible pitfalls that lie ahead.[67]

The R&D strength of China's business sector raises questions about its ability to absorb technology that universities seek to transfer under a Bayh–Dole strategy. The 'absorptive capacity' of China's business sector is one of those issues, however, that tend to generate a fair amount of debate, but no clear answer. There is plenty of anecdotal evidence to suggest that China's business sector is not as 'high-tech' as many developed-country business sectors. At the same time, China's evolutionary record over the last three decades demonstrates a highly adaptable business sector that responds to the incentive environment in which it operates. As the business parameters evolve in China to higher levels of technology – which appears to be what is occurring – there is little reason to believe that China's companies will not evolve as well. One part of China's business sector that could become a more significant technology transfer partner for China universities is its growing segment of high-technology start-up firms. Chapter 7 considers the future of China's start-up industry.

5.   PATENT ACTIVITY

Critics have long charged China with having weak IP protection. Because a Bayh–Dole strategy is predicated on universities transferring patent rights to the business sector, a weak IP protection scheme would be highly problematic because weak IP protection erodes the value of patents. Chapter 5 will analyze China's IP protection system in detail. This section will examine the substantial increase in patent activity that is taking place in China. While China's patent statistics need to be interpreted carefully, as the favorable data are accompanied by important caveats, the overall picture strongly suggests that patents are valuable property in China.

A patent provides its holder with an exclusive property right to exploit[68] an invention for some limited period of time (e.g., 20 years in the United States). In exchange for this potential exclusive right, a patent applicant must publicly disclose its invention. In theory, the patent process is designed to encourage technological development by providing a valuable incentive to inventors (through the exclusive property right), while at the same time improving society's overall level of knowledge (through the public disclosure requirement). Not surprisingly, countries that generate the most patent activity tend to be the more technology-developed countries, such as the United States, Japan and Germany. (But see Box 3.2 for a cautionary note on why patent statistics are not a perfect indicator of technology development.)

5.1   China's Patent Law Provides for Three Categories of Patents

China's patent system provides for three types of patents (a more detailed description of China's patent law is provided in Chapter 5):[69] invention patents; utility patents; and design patents.[70] In China, invention patents tend to be the most valuable of the three categories of patents. They tend to involve the most technologically advanced inventions, and therefore provide the inventor with the strongest property rights. Invention patents receive 20 years of protection, compared to 10 years for utility and design patents,[71] and involve a more thorough review process. When examining China's various patent statistics, it is important to differentiate between the more valuable invention patents, and the less valuable utility and design patents.

5.2   Substantial Increase in Patent Applications

China has experienced a huge increase in patent applications – in particular domestic patent applications – over the last decade (see Figure 3.7). The increase was most pronounced beginning in 2000. The increase in patent applications indicates a number of positive things about China's technology development. To the extent that the increased applications are being sought by Chinese inventors, it helps to show that creating technological innovations with a potential for commercial application has become part of the R&D culture in China. Figure 3.8 clearly shows that domestic applications have been the primary driver of China's

BOX 3.2 PATENT STATISTICS ARE NOT A PERFECT INDICATOR OF TECHNOLOGY DEVELOPMENT

While useful for understanding a country's inventive activity, a country's patent statistics are unlikely to be a perfect indicator. There are a number of reasons why patent statistics alone may provide a distorted picture of a country's inventive activity, including:[1]

- Patents are not the only technique for generating value out of an invention. The inventor, for example, may decide to rely on trade secret protection rather than a patent.
- The rights granted by the country's patent system may substantially influence patent activity. For example, the definition of what is 'patentable' can have a huge impact on patent activity (e.g., software patents or business-method patents).
- The mechanical operation of the patent system can also significantly influence the inventor's IP management strategy. For example, a low cost patent system (low application and maintenance fees) may encourage a substantial increase in the quantity of patents by making the decision to patent an easy one – even for inventors of lower-quality inventions with little commercial promise – but not necessarily increase the quality.
- It can be difficult to ascertain the 'quality' of patents from the patent activity statistics.
- R&D has not been immune to the outsourcing movement. R&D may be conducted in one country, but the commercialization of the invention (and therefore the patent protection) may occur in a different country.
- Patents tend to be a lagging indicator of a country's inventive activity. It takes years after a country has improved its inventive capabilities for patents to be filed, and then even more time for the patents to be granted.

Notes:
1. *See* WIPO (2008), *2008 World Patent Report – A Statistical Review*, p.10, available at www.wipo.int/ipstats/en/statistics/patents/wipo_pub_931.html.

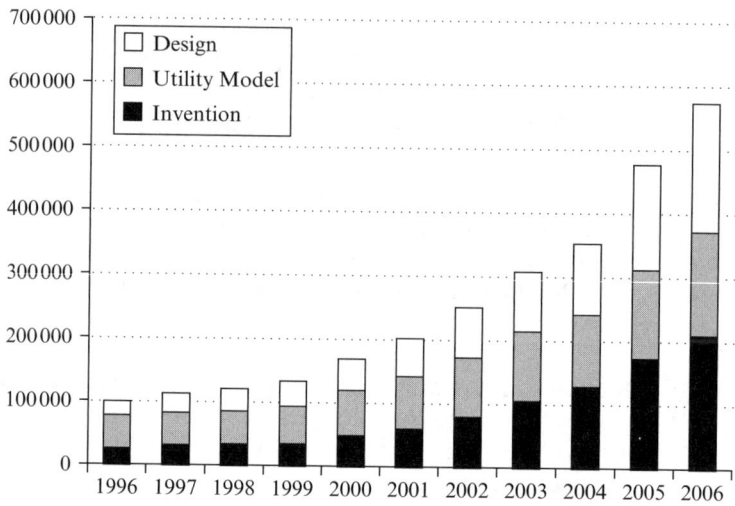

Source: China SIPO.

Figure 3.7 *Total patent applications received by China SIPO (1996–2006)*

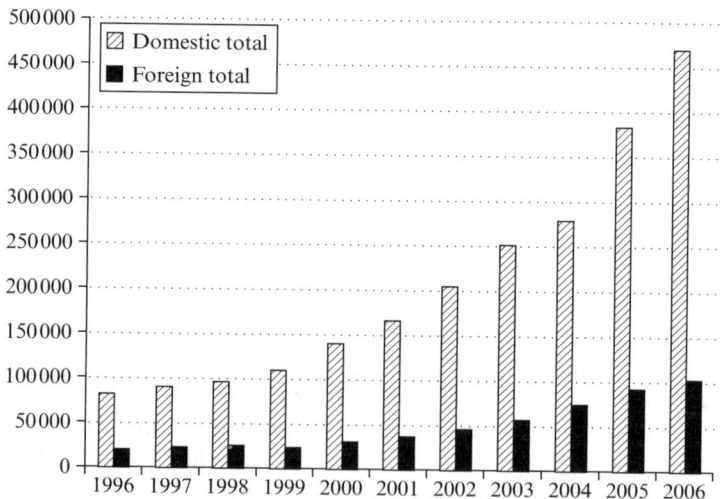

Source: China SIPO.

Figure 3.8 *Domestic v. foreign 'total' patent applications received by China SIPO (1996–2006)*

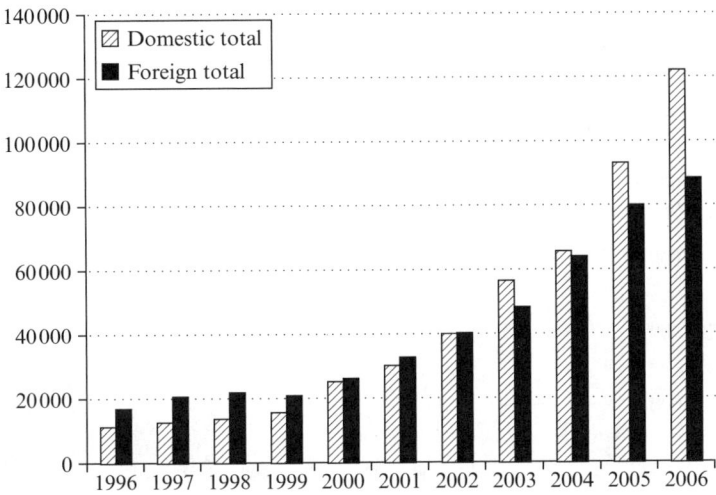

Source: China SIPO.

Figure 3.9   *Domestic v. foreign 'invention' patent applications received by China SIPO (1996–2006)*

increased patent applications. The number of domestic patent applications increased roughly fivefold from 1996 (82 207 total domestic patent applications) to 2006 (470 342 total domestic patent applications). While the primary driver of China's increased patent applications was domestic inventors, the level of foreign applications has also increased substantially since 1996 (20 528 total foreign applications in 1996 compared to 102 836 in 2006). This increase in foreign applications helps to show China's increased relevance as an overall technology market with more foreign inventors feeling the need to seek patent protection in China.

The above patent application statistics provide a relatively positive picture of China's R&D output, but there is a general concern expressed by many that China's domestic patent success has been one of quantity, and not quality. Chinese inventors are generating lots of patent applications, but the overall quality is low. The available patent statistics lend some support to that notion. Figure 3.9 shows that while Chinese inventors may be the dominant source of patent applications generally in China, they do not dominate the applications for invention patents, and have only recently begun to file more invention patent applications than foreign inventors.

Moreover, the mechanics of the Chinese patent filing system may be

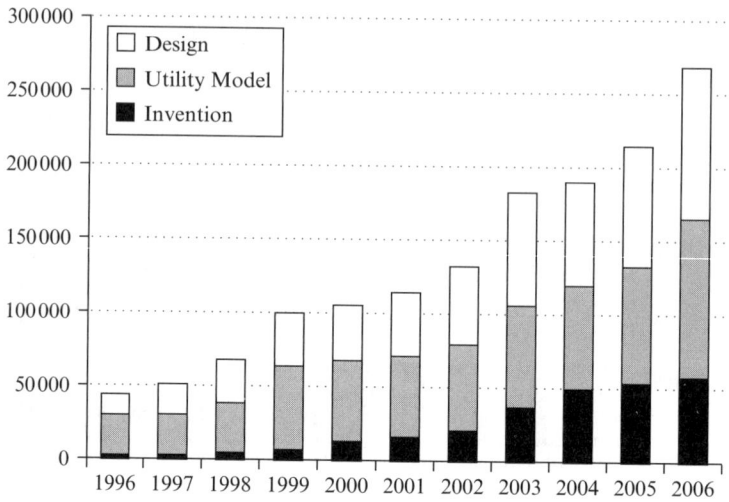

Source: China SIPO.

Figure 3.10   *Total patents granted by China SIPO (1996–2006)*

encouraging some double counting of patent applications. Because (i) patent applications are relatively inexpensive in China, (ii) invention patents go through a longer review process, and (iii) there is no prohibition against applying for both an invention patent and a design patent for the same invention, some Chinese inventors employ a double-patent-application strategy, namely, the inventor applies for both an invention patent and a design patent for the same invention in order to shorten the patent pending period.[72] While SIPO will not issue 'double patents' – i.e., an inventor cannot obtain a design patent and an invention patent on the same invention[73] – it is not clear whether SIPO is able to filter out the double patent filing strategy from its 'application statistics.' As a result, it could be that patent application figures have been falsely inflated by double-counting some of the patent applications.

5.3   Substantial Increase in Patents Granted

Patent application statistics provide an indication of inventors' efforts, but they provide little insight into the quality of those efforts. Patent grant statistics are somewhat more useful for that analysis. As with patent applications, the number of patents granted by China SIPO has greatly increased over the last decade (see Figure 3.10). Because the patent approval process

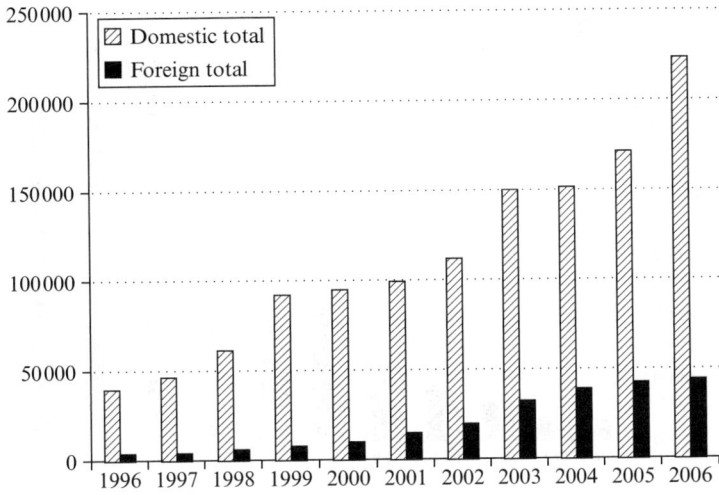

Source: China SIPO.

Figure 3.11 *Domestic v. foreign 'total' patents granted by China SIPO (1996–2006)*

takes a few years, the big jump in patents granted took place in 2003, a few years after the big patent application increase in 2000.

Figure 3.11 clearly shows the vast majority of the increase stems from patents granted to Chinese inventors. The number of domestic patents granted increased roughly fivefold from 1996 (39 725 total domestic patents granted) to 2006 (223 860 total domestic patents granted).

As with the patent application statistics, the patent grant statistics are not all favorable and many of the statistics support the notion that Chinese inventors are generating large quantities of lower quality inventions (see Figure 3.12). For domestic inventors, only 8% of patents granted between 1996 and 2006 were for invention patents (100 534 out of 1 242 605 total domestic patents granted). That ratio has been improving recently, but invention patents remain a relatively small percentage of domestic patents granted in China. From 2003 to 2006, for example, 11% of patents granted to domestic inventors were for invention patents (75 427 out of 696 395 total domestic patents granted). By comparison, foreign inventors generated more invention patents from 2003 to 2006 (foreign inventors received 122 178 invention patents during that period) and invention patents accounted for 77% of patents granted to foreign inventors (122 178 out of 158 074 total foreign patents granted).

Shaping China's innovation future

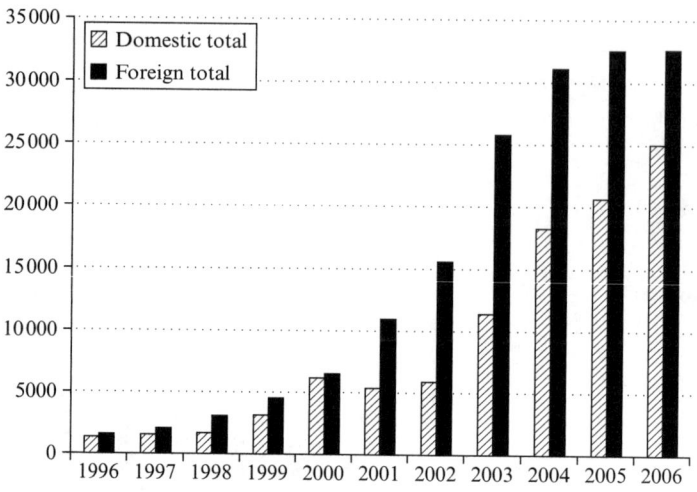

Source: China SIPO.

Figure 3.12   Domestic v. foreign 'invention' patents granted by China SIPO (1996–2006)

5.4   Analysis of 'Patent Families'

Another way to analyze the quality of China's domestic patents is to examine the inventors' international patent strategy. A Chinese patent offers the inventor protection in China, but not outside China. For example, a Chinese inventor that is granted a patent by China SIPO does not receive any protection from infringement that may take place in the United States, Japan, or Europe. The highest quality inventions, however, should have global commercial appeal. To take advantage of that global commercial appeal, the inventor will need to pursue a global patent strategy. The inventor must apply for and receive a patent in each country of interest in its global commercial strategy. Presumably, 'patent families' – a set of interrelated patent applications for a related invention in multiple countries[74] – are therefore more valuable than a purely domestic patent. China has made substantial progress in its growth of patent families, but still remains far behind the more developed countries. Table 3.14 shows China's progress in generating U.S. patent grants.

A recent OECD report on 'triadic patent families' – a set of patent applications taken at the European Patent Office, the Japan Patent Office and the U.S. Patent Office – found the following about China's triadic patent families:[75]

Table 3.14   U.S. patents granted by country – all patent types

Residence of first-named inventor	1987	1992	1997	2002	2007
United States	47916	58791	69923	97125	93691
Japan	17294	23164	24191	36339	35942
Germany	8093	7605	7292	11957	10012
Taiwan	411	1253	2597	6730	7491
South Korea	105	586	1965	4009	7264
Canada	1770	2231	2817	3857	3970
France	2990	3282	3202	4421	3720
China	23	41	66	390	1235
India	12	24	48	267	578
Brazil	35	43	67	112	118

Source: USPTO (2009), *Patent Technology Monitoring Team Report, Patent Counts by Country/State and Year all Patents, All Types: January 1, 1977 – December 31, 2008*, available at www.uspto.gov/web/offices/ac/ido/oeip/taf/cst_all.pdf.

- China has experienced average annual growth of 33% in its triadic patent families between 1995 and 2005. The usefulness of this growth number, however, is questionable, given that China was starting from almost zero in 1995.
- On a total number basis, China became one of the top 15 patenting countries in 2005.
- On a *per capita* basis, China has one of the lowest international patenting propensities, with fewer than 0.3 triadic patent families per million people – although that number is growing rapidly.

All told, China's patent statistics demonstrate that China is making progress in its technological development. It is generating more applications and granting more patents, but for the time being China's patent success appears to be much more about quantity than quality.

6.   INTERACTIONS AMONG THE KEY TECHNOLOGY ACTORS

6.1   Dynamic Nature of Innovation Systems

One reason that innovation systems are so difficult to analyze, and therefore thoughtfully improve, is their incredibly dynamic nature. How does

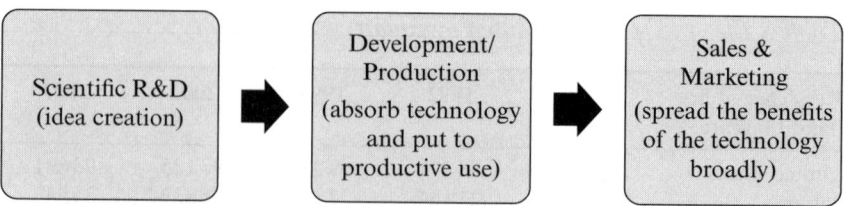

Figure 3.13   Linear vision of an innovation system

an innovation system function? If you ask most participants in the innovation system – whether it be university scientists, private sector engineers, or politicians – they tend to envision the innovation system as functioning in a relatively linear fashion. A typical description might go something like this:

1. Innovation starts with scientific R&D.
2. As valuable new technologies are discovered through the R&D process, the business sector takes those technologies and develops them into workable products and services.
3. These marketable products and services are then sold to the general public.

The flow of knowledge is quite easy to follow in such a linear world (see Figure 3.13).

It would be wonderful if innovation systems were that simple. If that were the case, countries wishing to increase the flow of technology through their economy could simply focus on increasing R&D spending (in particular basic research spending), and the resulting discoveries would flow through the linear pipeline and generate more technology-based products, processes and services, and more robust technology-based industries. Unfortunately, innovations systems are nowhere near that simple. Innovation systems are inherently messy affairs, with an almost endless list of allocation decisions that must be made by the various innovation actors on a constant basis. Where should researchers focus their research? How does the research sector coordinate its activities with the business sector? What happens if the business sector is not capable of absorbing the technology being developed from the R&D processes? What happens if the R&D processes are not capable of generating the technology needed by the business sector and/or demanded by the general public? Which technology endeavors should be funded? For those that are funded, how much funding should they receive? What role does financing start-ups play in an innovation system? What if the government fails to understand the innovation process adequately and has a penchant for adopting bad policies that hurt the process? Linear thinking about innovation systems makes it difficult to answer any of these questions, even though they are all very common occurrences in any innovation system.

When viewed as a complex set of relationships between the various actors,[76] however, the description of how an innovation system actually works becomes a lot more accurate and these types of questions, as well as many others, become much easier to address. The transfer of university-developed technology to the business sector, for example, takes place within an innovation system. The ability for universities to contribute to their economies by consistently commercializing technology is an inherently complex endeavor that involves a matrix of interconnected components, including

- Talented university scientists;
- Industry that is capable of absorbing the developed technology and informing future developments;
- A highly-skilled workforce that is capable of operating in an innovation-based environment;
- Adequate funding for university research;
- Adequate funding for start-up and other high-technology companies to develop businesses based on university-developed technology;
- University administrators that build institutions within the university to support the commercialization process and create a technology commercialization culture; and
- An institutional framework (both formal and informal) that facilitates productive exchanges between these various parties.

University-based innovation systems that provide better linkages and coordination between these various actors and activities should be much more efficient and productive than university innovation systems with weaker linkages and coordination. While extremely valuable, these linkages can be very difficult to establish and do not develop as frequently, or as easily, as one might expect based on their usefulness. The link between university scientists and industry is a perfect example of this concept.

6.2 University/Industry Linkages

The legacy of more than 30 years of a 'functional specialization' approach to innovation has not yet been completely eliminated in China. Fortunately,

Table 3.15   R&D cross-funding across sectors (in billions of RMB)

Source of funds	2004		2005		2006		2007	
	Amount	%	Amount	%	Amount	%	Amount	%
R&D funding for business sector								
Government	6.3	4.8	7.7	4.6	9.7	4.5	12.9	4.8
Business	118.9	90.5	152.7	91.2	194.6	91.2	246.6	91.9
Sources from abroad	2.0	1.5	1.7	1.0	4.2	2.0	4.2	1.6
Other	4.2	3.2	5.3	3.2	5.0	2.3	4.6	1.7
R&D funding for GRIs								
Government	34.4	79.8	42.5	82.8	48.1	84.8	59.3	86.2
Business	2.2	5.2	1.8	3.4	1.7	3.1	2.6	3.8
Sources from abroad	0.3	0.6	0.2	0.4	0.3	0.5	0.3	0.5
Other	6.2	14.5	6.9	13.5	6.6	11.7	6.5	9.5
R&D funding for higher education sector								
Government	10.9	54.2	13.3	55.0	15.2	54.9	17.8	56.5
Business	7.5	37.3	8.9	36.8	10.1	36.5	11.0	34.9
Sources from abroad	0.3	1.5	–	–	0.4	1.4	0.5	1.6
Other	1.5	7.5	0.2	0.8%	2.0	7.2	2.2	7.0

Source: MOST (2006–2008), *China S&T Statistics Databooks.*

however, productive linkages between China's core technology actors are being formed, including between its university and business sectors. The remainder of this chapter will explore certain of these university/industry linkages.

6.2.1   Cross funding

Each of China's primary R&D actors receives R&D funding from a variety of sources (see Table 3.15). This cross-funding, in particular between the university and business sectors, helps build valuable linkages between the actors.

Table 3.15 provides some potentially interesting insights about R&D funding in China and its role in fostering science/industry linkages. GRIs are the biggest recipients of government funding, and receive only minor support from the business sector. With the 1999/2000 restructuring of GRIs to focus on basic and applied research, and on research with a

Table 3.16   Level of R&D cooperation between large and medium-sized Chinese companies and the higher education/GRI sectors

Co-operation pattern in R&D projects by large and medium-sized Chinese companies	2000	2003	2005
% of R&D projects pursued independently by the Chinese company	70.8%	73.5%	77.7%
% of R&D projects that involved cooperation with a Chinese higher education institution	8.0%	8.5%	4.4%
% of R&D projects that involved cooperation with a GRI	7.6%	7.3%	9.7%
Total number of R&D projects in the survey	**23 576**	**24 665**	**39 072**

Source: Martin Schaaper, 'Measuring China's Innovation System: National Specificities and International Comparisons', *OECD Science, Technology and Industry Working Papers* 17, 29 (2009), available at www.oecd.org/dataoecd/15/55/42003188.pdf (citing *China Statistical Yearbook on Science and Technology* (2006) Table 2.27).

'public-goods quality' like agriculture and defense, the level of government funding is not surprising. The government is the most logical source of funds for that type of research. The higher education sector, however, receives more than a third of its R&D funding from the business sector, which is exactly the type of linkage one hopes to see in a successful Bayh–Dole environment.

6.2.2   Co-operation in R&D projects

Another indication of the level of linkage between the business sector and universities is the level of their cooperation in R&D projects. Table 3.16 provides data from a 2006 OECD report on China's innovation system.[77] The data from Table 3.16 came from surveys of R&D projects by large and medium-sized Chinese companies.[78] The level of co-operation between enterprises and universities is very low, and dropped (on a percentage basis) substantially in 2005. The OECD report explains the need for improving this cooperation:

At present, less than one-third of large and medium-sized industrial enterprises have their own research units, indicating the wide-spread difficulty for these enterprises to rely on [their] own R&D to cater their needs for technology and innovation. Therefore, China's large and medium-sized industrial enterprises need to develop science–industry linkages to utilise the R&D resources of the higher education sector and research institutes and to enhance [their] R&D capacity through co-operation and technology diffusion.[79]

Table 3.17   R&D outsourcing by China's business sector

	2000	2004
Large Enterprises		
Number of firms	1427	2136
Number of S&T firms	1180	1592
Intramural S&T expenditures	44.3 billion RMB	121.5 billion RMB
Extramural S&T expenditures	5.2 billion RMB	11.5 billion RMB
● To Chinese GRIs and Universities	2.5 billion RMB	3.8 billion RMB
Medium Enterprises		
Number of firms	17 680	25 574
Number of S&T firms	7832	9034
Intramural S&T expenditures	38.4 billion RMB	78.5 billion RMB
Extramural S&T expenditures	4.5 billion RMB	8.0 billion RMB
● To Chinese GRIs and Universities	1.5 billion RMB	2.4 billion RMB
Small Enterprises		
Number of firms	143 785	248 813
Number of S&T firms	15 125	22 307
Intramural S&T expenditures	18.2 billion RMB	40.0 billion RMB
Extramural S&T expenditures	1.4 billion RMB	1.8 billion RMB
● To Chinese GRIs and Universities	0.7 billion RMB	0.9 billion RMB

Source: Martin Schaaper (2009), 'Measuring China's Innovation System: National Specificities and International Comparisons', *OECD Science, Technology and Industry Working Papers* 17, 29 (2009), available at www.oecd.org/dataoecd/15/55/42003188.pdf (citing to microdata estimates, National Bureau of Statistics, as source for table).

6.2.3   R&D outsourcing

The 2009 OECD report on China's innovation system also presented data on the R&D outsourcing activities of China's business sector (see Table 3.17). While outsourced R&D remains a small portion of China's business sector R&D expenditures, Chinese enterprises have demonstrated some willingness to outsource R&D dollars to China's GRIs and universities.

6.2.4   University science parks

In the early 2000s, the creation of university-based science parks developed as a trend. The Ministry of Science and Technology and the Ministry of Education certified 22 'national-level university science parks' in 2001 and another 21 in 2002.[80] As of 2009, there were 69 such university science parks in China serving 109 universities (see Table 3.18). A number of these university-based science parks are located within one of the Torch

Table 3.18   National-level university science parks

Park	Universities	Park	Universities
13 parks located in Beijing			
Beijing Institute of Technology Science Park	Beijing Institute of Technology	China Agricultural University Science Park	China Agricultural University
Beijing Jiaotong University Science Park	Beijing Jiaotong University	North China Electric Power University Science Park	North China Electric Power University
Beijing University of Areonautics & Astronautics Science Park	Beijing University of Areonautics & Astronautics	Peking University Science Park	Peking University
Beijing University of Chemical Technology Science Park	Beijing University of Chemical Technology	Renmin University of China Science Park	Renmin University of China Science Park
Beijing University of Posts and Telecom-munications Science Park	Beijing University of Posts and Telecommunications	Tsinghua University Science Park	Tsinghua University
Beijing University of Technology Science Park	Beijing University of Technology	University of Science & Technology Beijing Science Park	University of Science & Technology Beijing
BNU-BUCM Science Park	Beijing Normal University; Beijing University of Chinese Medicine		
10 parks located in Shanghai			
Donghua University Science Park	Donghua University	Shanghai Jiaotong University Science Park	Shanghai Jiaotong University
East China Normal University Science Park	East China Normal University	Shanghai University Science Park	Shanghai University
East China University of Science and Technology Science Park	East China University of Science and Technology	Shanghai University of Finance and Economics Science Park	Shanghai University of Finance and Economics
Fudan University Science Park	Fudan University	Tongji University Science Park	Tongji University Science Park
Shanghai Institute of Electric Power Science Park	Shanghai Institute of Electric Power	University of Shanghai for Science and Technology Science Park	University of Shanghai for Science and Technology

Table 3.18 (continued)

Park	Universities	Park	Universities
7 parks located in Jiangsu			
Changzhou Science Park	Nanjing University Changzhou High-Tech Institute, Changzhou Academe of Southeast University, Jiangsu Polytechnic University, Jiangsu Teachers University of Technology, Hehai University (Changzhou), Changzhou Institute of Technology, Changzhou College of Information Technology, Changzhou Institute of Engineering Technology, Changzhou Institute of Light Industry Technology, Changzhou Institute of Mechatronic Technology, Changzhou Textile Garment Institute	Science Park of Nanjing University and Colleges in Gulou District	Nanjing University, The Government of Gulou District, Nanjing University of Posts and Telecommunications, Hehai University, Nanjing Institute of Technology, Nanjing University of Chinese Medicine, Nanjing Medical University, Nanjing Normal University, China Pharmaceutical University
China University of Mining and Technology Science Park	China University of Mining and Technology (Xuzhou)	Southeast University Science Park	Southeast University
Jiangnan University Science Park	Jiangnan University		
Nanjing University of Science and Technology Science Park	Nanjing University of Science and Technology		
Nanjing University of Technology Science Park	Nanjing University of Technology		

Park	Universities	Park	Universities
5 parks located in Shaanxi			
Northwest A&F University Science Park	Northwest A&F University	Xi'an Jiaotong University Science Park	Xi'an Jiaotong University
Northwestern Polytechnical University Science Park	Northwestern Polytechnical University	Xidian University Science Park	Xidian University
Shanxi North University of China Science Park	North University of China		
4 parks located in Sichuan			
Sichuan University Science Park	Sichuan University	Southwest University of Science and Technology Science Park	Southwest University of Science and Technology
Southwest Jiaotong University Science Park	Southwest Jiaotong University	University of Electronic Science and Technology of China Science Park	University of Electronic Science and Technology of China
3 parks located in Guangdong			
Shenzhen Virtual University Park	Shenzhen High-Tech Industrial Park	Sun Yat-Sen University Science Park	Sun Yat-Sen University
South China University of Technology Science Park	South China University of Technology		
3 parks located in Liaoning			
Dalian University of Technology/ Qixianling Science Park	Dalian University of Technology, Dalian Maritime University, Dalian Medical University, Dalian Fisheries University, Dongbei University of Finance & Economics, Dalian Institute of Chemical Physics (DICP) under Chinese Academy of Sciences (CAS)	Shenyang University of Technology Science Park	Shenyang University of Technology

Table 3.18   (continued)

Park	Universities	Park	Universities
Northeastern University Science Park	Northeastern University		
	3 parks located in Heilongjiang		
Harbin Engineering University Science Park	Harbin Engineering University	Harbin University of Science and Technology Science Park	Harbin University of Science and Technology
Harbin Institute of Technology Science Park	Harbin Institute of Technology		
	3 parks located in Tianjin		
Hebei University of Technology Science Park	Hebei University of Technology	Tianjin University Science Park	Tianjin University
Nankai University Science Park	Nankai University		
	2 parks located in Chongqing		
Chongqing Beibei Science Park	Southwest Normal University, Southwest Agricultural University	Chongqing University Science Park	Chongqing University
	2 parks located in Gansu		
Lanzhou Jiaotong University Science Park	Lanzhou Jiaotong University	Lanzhou University Science Park	Lanzhou University
	2 parks located in Shandong		
China University of Petroleum Science Park	China University of Petroleum	Shandong University Science Park	Shandong University
	2 parks located in Zhejiang		
Science Park of Zhejiang Province	Zhejiang University of Technology, Zhejiang Sci-Tech University, China Jiliang University, The Government of Jianggan District	Zhejiang University Science Park	Zhejiang University

Park	Universities	Park	Universities
1 park located in each of the following provinces:			
Anhui, Fujian, Hebei, Hubei, Henan, Hunan, Jiangxi, Jilin, Xinjiang, Yunan			
East Lake High-Tech Zone Science Park (Hubei)	Huazhong University of Science & Technology, Wuhan University, Huazhong Agricultural University, Wuhan University of Technology, Huazhong Normal University	Science & Technology Park of Yunnan Province (Yunnan)	Yunnan University, Kunming University of Science and Technology, Yunnan Normal University, Yunnan Agricultural University, Kunming Medical University, Dali University, Yunnan University of Finance & Economics
Hefei Science Park (Anhui)	University of Science and Technology of China, HeFei University of Technology, Anhui University	Xiamen University Science Park (Fujian)	Xiamen University
Jilin University Harbin Institute of Technology Science Park (Jilin)	Jilin University	Xinjiang University Science Park (Xinjiang)	Xinjiang University
Nanchang University Science Park (Jiangxi)	Nanchang University	Yanshan University Science Park (Hebei)	Yanshan University
Science Park of Henan Province (Henan)	Zhengzhou University, Henan Agricultural University, Henan University of Technology, Zhengzhou University of Light Industry	Yuelu Mountain Science and Technology Park (Hunan)	Central South University, Hunan University, National University of Defense Technology, Changsha Research Institute Of Mining And Metallurgy, Changsha Research Institute of Mining, Hunan Normal University, Hunan Academy of Chinese Medicine, Changsha Construction Machinery Research Institute

Source: Ministry of Education of the PRC, available at www.moe.gov.cn/edoas/website18/72/info1240803227649272.htm.

Program development zones.[81] For the most part, these university science parks perform classic 'incubator' functions. They provide a convenient location for start-ups to operate by placing the start-ups in close physical proximity to research universities (and their laboratories) and to other high-technology companies, which creates an environment where valuable technology transfer can take place. Moreover, by locating a critical mass of high-technology companies in one physical location, it also makes it easier for technology intermediaries such as venture capital firms to perform their function of financing and nurturing start-ups. While there are no official data on the economic output of these university science parks,[82] they do demonstrate that some level of physical linkage exists between China's universities and industry. In the future, these university science parks could serve as convenient vehicles for running experiments to determine which policies and procedures are most effective at encouraging efficient technology exchanges between universities and industry.

6.3 Other Important Linkages

This section has focused primarily on science/industry linkages. There are other linkages that play significant roles in the success of a country's Bayh–Dole strategy, such as:

- The link between the government and the science and business sectors. Does the government understand and support the innovation process?
- The link between private sources of capital and innovative companies seeking to grow. Are high-technology companies capable of raising capital to finance their ventures?

These linkages will be addressed in later chapters. Chapter 7 will examine the Chinese government's support of the innovation process, and Chapter 8 will explore the finance environment for innovative companies in China and the ability of venture capital firms to develop valuable linkages to help finance small- to medium-sized innovative companies in China.

PART II

The legal and policy environment for commercializing university technology in China

INTRODUCTORY NOTE TO PART II

A country's legal and policy environment plays a critical role in encouraging, or discouraging, the formal flow of commercial technology from universities to the business sector. Reduced to its most basic ingredients, successful university technology transfer requires universities with strong R&D capacity to create new technology and technology-based businesses that are capable of absorbing that technology. Developing those two basic ingredients and optimizing their interaction is highly dependent on the institutional environment in which they operate. China's development of a market-based innovation system is a prime example of the positive impact that institutional change can have on university technology transfer.

In addition to having a market-based innovation system (Part I of this book), there are five critically important policies that will bear on the success of China's university technology transfer system:

1. Does the country's legal system support market-based transactions in general (Chapter 4)?
2. Does the country's legal system specifically protect IP on a sufficient level (Chapter 5)?
3. Who owns the IP rights to government-funded, university inventions, and what rights are afforded to the owner (Chapter 6)?
4. Does the government dedicate sufficient public funds to university research, and are those funds allocated in an efficient manner (Chapter 7)?
5. Does the government provide a sufficiently supportive environment for innovative, technology-based companies so that universities have a sufficiently strong business sector with which to partner (Chapters 7 and 8)?

Part II will examine each of these issues in detail for China.

4. Developing a legal system that supports the market-based transactions of a Bayh–Dole strategy

> Let the laws roll with the times and there will be good government. . . But let the times shift without any alteration in the laws and there will be disorder.
> – Han Fei Tzu (d. 233 BC), chief theoretician of China's Legalist school[1]

Economic actors will invest in the creation, adoption and dissemination of technology if they believe they will receive a profitable return from that effort. Without that belief, it makes little sense for them to invest the time and effort required to undertake such burdensome activities. China's efforts to create a market-based innovation system would likely go for naught if the market reforms were not supported by a complementary effort to build expectations among its various actors that they can profit from participating in such a system. A country's legal system can play a major role in providing the expectations needed for economic actors to act productively. A legal system can create confidence in property ownership that incentivizes property owners to develop and employ their property in a productive manner. It can also create confidence in the enforceability of contracts, which substantially reduces the overall cost of economic exchanges. While a formal legal system is not the only technique that policymakers can employ to create the needed expectations,[2] it is the dominant approach in industrialized countries and has become a major point of focus in many developing countries.[3] One explanation for China's technology transfer failures during the 1980s – e.g., the mid-1980s technology market that was supposed to transfer GRI-developed technology to the business sector – was the weak state of China's legal system at the time.

China's current Bayh–Dole system is highly dependent on a robust legal system, as it is based on well-defined property rights in innovations and the legal transfer of such rights through enforceable contracts. China's Bayh–Dole system is very much a 'law-based' strategy. China's legal system has made considerable progress since the failed technology market, and this chapter explores whether it has progressed enough to support the market-based transactions that are central to a Bayh–Dole strategy.

Deciding whether to include this chapter (and, if so, where to place it) was one of the more difficult decisions we encountered in writing this book. On the one hand, there were a number of valid arguments for not including the chapter. In developed countries, little thought is given to whether a country's legal system generally supports market-based transactions, as the legal system has likely performed such a role for centuries. As a result, the legal analysis of a country's university technology transfer capacity will invariably focus on the sufficiency of the country's IP laws and who owns the IP rights to government-funded, university inventions. Seldom does the analysis include a look at the general sufficiency of the legal system, so this chapter may seem like a distraction for some readers. Moreover, the chapter poses problems with the logical flow of the book. The subsequent chapters on China's IP laws (Chapter 5) and China's Bayh–Dole System (Chapter 6) have a natural connection with Chapter 3's analysis of China's current innovation system.

On the other hand, we have many occasions to speak with non-Chinese businesspersons and lawyers about doing business in China, and their perceptions of China's legal system are not too complimentary. Far too often, we hear from these non-Chinese actors their belief that China does not have a 'real' legal system. When talking about whether China's legal system is sufficient to support the flow of publicly-funded university innovations into the commercial sector, we are often met with quizzical looks and questions such as:

- How can you talk about a legal strategy for university technology if China does not respect the rule of law?
- Bayh–Dole strategies involve highly-sophisticated and complex legal concepts. Are China's lawyers and judges competent to oversee such strategies?
- Are private property rights really available in China, and can you really rely on contracts to transfer property rights?

While our desire may be to just let out a deep sigh when we hear such questions – since China's legal system, although imperfect, is more than sufficient to support the market-based transactions of a Bayh–Dole strategy – we have found that the better course is to help educate people about the strength and quality of China's legal system. In addition, foreign businesses have been, and will continue to be, major acquirers of the technology developed by Chinese universities. Unfairly negative perceptions about the quality of China's legal system can inappropriately dissuade foreign partners from acquiring Chinese university technology. In the end, we feel it is important to combat outdated perceptions of China's legal system and we decided to include this chapter. Those readers who are familiar with China's legal system and the incredible progress it has made since the outset of the reform era in 1978 can skip to Chapter 5. For those readers who want to learn about the creation of a modern, market-supporting legal system, basically from scratch, and how that legal system has become thoroughly integrated into economic transactions in China, then Chapter 4 should be worth a quick detour before moving on to China's IP and Bayh–Dole systems.

1. NOT THAT LONG AGO, CHINA'S ECONOMIC LEGAL SYSTEM DESERVED ITS POOR REPUTATION – SOME HISTORICAL PERSPECTIVES

At the start of China's reform period, in 1978, the legal infrastructure needed for a market-driven economy was simply non-existent in China.[4] At that time, there were almost no lawyers in China, there was only a minimal amount of civil law on the books and the existing court system was not capable of handling commercial disputes. Lawyers had largely disappeared from China by the start of the reform period. Law schools were shut down early in the Cultural Revolution and few had reopened before 1978.[5] China's pre-Mao nationalist civil codes were eliminated following the establishment of the People's Republic of China, and had not yet been replaced. The Cultural Revolution also wreaked havoc on the court system. During the Cultural Revolution, more traditional courts disappeared and were replaced by a variety of dispute resolution mechanisms, including various Party organs, the police, street committees and rural peoples' communes.[6] All told, China had to create an entire legal infrastructure, almost from scratch, beginning in 1978. Civil codes had to be adopted; courts created; lawyers and judges had to be trained; and legal processes, procedures and norms of behavior had to develop.

Since the start of the reform era, however, China's legal system has shown a tendency to evolve in tandem with the country's economy. Chinese legal scholar Donald Clarke describes the relationship between legal and economic development in China as 'bidirectional – a co-evolutionary process.'[7] Early economic reforms, for example, allowed for the gradual development of new legal actors within the economy (e.g., Town and Village Enterprises, or TVEs) and new economic activities that did not naturally fall under China's then-existing supervision institutions.

A new set of rules and institutions was needed to legitimize and organize these new actors and their new economic activities, and the law eventually developed to a stage where it could fill this role.[8] As the law grew to fill this role, it began to provide the expectations needed for China's economic actors to act productively, which supported further economic development and encouraged China's policymakers to experiment with even greater economic liberalization – which required further legal reform.[9]

To comprehend this co-evolutionary process, it is useful to have a general understanding of China's economic reform movement so as to better appreciate the context in which the law developed.

1.1 Early Reform Period (1978–1992)

China's market transition officially began in December 1978 with the Third Plenary Session of the 11th Central Committee of the Communist Party of China.[10] The Third Plenary Session was a historic event in China that marked the return to power of Deng Xiaoping and the commencement in earnest of China's reform period.[11] The Central Committee declared an end to 'protracted class struggle' as the party's central focus and replaced it with the Four Modernizations[12] – i.e., the modernization of China's agricultural, defense, industrial and S&T sectors. In effect, the party replaced its prior focus on political and ideological goals with a particular focus on economic development.[13]

A host of new policies and economic reforms followed the Third Plenary Session. At the outset of the reform period, 'the extent of the possible was not known, and experimental reforms were launched in nearly every sector of the economy.'[14] While the reforms were widespread and varied, a few broad generalizations can be made about the early phases of China's market transition. The most successful of the early reforms took place in the countryside with the farming sector.[15] The rural sector reforms – e.g., initiating the household responsibility system in agriculture, which allowed farmers to sell surplus crops through a market process, and the establishment of TVEs – were the most market-based of China's early economic reforms.

Like the initial reforms to China's innovation system, most of the early economic reforms (other than the rural reforms) were primarily aimed at improving China's plan-based system, not developing a market-based system.[16] Over time, China's leadership abandoned its effort to 'fix' the old planned-based system and gradually increased its focus on market-based reforms.[17] During the 1980s, a gradual relaxation of the state's monopoly over industry coupled with a gradual relaxation of price controls (i.e., China's dual-track pricing system) were the most fundamental of these market-based reforms. New firms entered the economy during the 1980s that were able to compete due to the gradual decontrol of prices.[18] At the same time as the non-state sector was developing, China's government was also trying to improve the management and performance of the state sector. In spite of these efforts, the new non-state sector grew faster than China's state sector.[19] Eventually, the growth of China's non-state sector caused China to 'grow out of the plan' – a phrase used by Barry Naughton, a noted expert on China's economy, to describe how non-state economic growth gradually eroded the importance of central planning and the state sector.[20]

While China's market reforms eventually took hold in the 1980s and proved to be successful, it was not always clear during that time that developing a market economy was an inevitable outcome. Market reforms were constantly scrutinized and subject to harsh criticism from conservatives who were not convinced of the wisdom of the market-based reforms.[21] This policy competition led to a reform pattern that is sometimes referred to as 'two steps forward, one step back.'[22] Market-based reforms were strong in some years (i.e., 1979, 1984, 1987–1988), but seemed to retreat in others (i.e., 1981–1982, 1986, 1989).[23]

1.2 The Early Role of Law in Post-Reform China

During the early phases of the reform era, the law played a relatively minor role in China's economic activities. Because there was no abrupt shift to a market economy, there was no need for the law to immediately take on a central role in China's economy. The state sector remained the dominant force in China's economy, and the state sector had already been operating for roughly 30 years without much assistance from a formal legal system. Just as the continuation of the state sector allowed China's market reforms to gradually develop and 'grow out of the plan,' the continuation of a dominant state sector also allowed for a gradual increase in the significance of law in China's economic realm.[24] That is not to say that the law was irrelevant, or not viewed by policymakers as an important potential tool for future economic development, but its role in day-to-day economic activities was simply not that extensive at the outset of the reform era.[25]

1.2.1 Legalization

The law's initial, reform-era role had little to do with market reforms, but instead focused on improving the administration of the state sector. At the outset of the reform era, the state's bureaucracy was extremely inefficient. Bureaucrats did not take their orders from uniform, impersonal laws or administrative regulations that were free for all to see. Instead, direc-tives tended to be made through internal bureaucratic communications (that were not uniformly applied and often lacked clear authority),[26] and general policy pronouncements that were drafted in such general terms that their ability to provide useful and practical guidance to bureaucrats was highly questionable.[27] The result was a system that was 'incapable of imposing unity and order upon the processes of government.'[28] The law's earliest function was to provide improved order to this system.

The law performed a similar function in the economic sector, and was intended to replace the 'particularistic bargaining regime of the past with a regime of strict, impersonal, and universalistic rules that would impose discipline . . . and encourage efficiency.'[29] In effect, the law was used to *legalize* various government administration processes and the operation of the state sector.[30] Over time, promulgated laws and regulations have replaced much of the prior informal governance structure. Professors Clarke, Murrell and Whiting use the example of China's Bankruptcy Law of 1986 to illustrate this *legalization* mechanism:

> A prime example of this type of mechanism was the Bankruptcy Law (1986), passed in 1986. According to contemporary commentary, '[T]he threat of bankruptcy urges all enterprises and people on and will turn muddleheaded people into shrewd ones and lazy people into diligent ones.'[31] . . . Prior to the Bankruptcy Law, however, the state already had the power to close down loss-making enterprises. Thus this law and similar enactments were as much efforts by the state to effect policy in a new way as they were new policies.[32]

1.2.2 Governing the interactions of independent economic agents

Over time, the role of law began to evolve in China past the simple *legali-zation* function to play the more traditional economic role of defining the rules for interactions between independent economic actors. As China's non-state sector grew in importance during the 1980s, the need for a formal legal system to legitimize (and eventually facilitate) its activities also grew. This transition to a more modern legal system took hold very gradually throughout the 1980s. The development of China's contract and business association laws during that period helps to illustrate this evolv-ing role of law in China.

a. Contract law   China's initial reform-era contact law – the Economic Contract Law (1981) – was adopted in 1981 and is probably best char-acterized as an example of the *legalization* function. One of the most notable features of the Economic Contract Law was its Article 2, which restricted the ability to contract to 'legal persons'[33] (e.g., enterprises or eco-nomic units).[34] Presumably, one of the major purposes of the Economic

Contract Law was to increase uniformity and transparency in government allocation decisions by replacing bureaucratic directives with economic contracts. Individuals, as well as entities that were not yet recognized by China's legal system, were not given authority to contract, which had the effect of excluding them from China's more meaningful economic activity.[35] That problem was short-lived because by the mid-1980s, China was expanding the scope of enforceable contracts to recognize 'non-state enterprises as independent economic actors'[36] and to establish formal rules for Sino-foreign contracts.[37] In 1986, for example, China adopted the General Principles of Civil Law (GPCL), which partially codified 'the lawful civil rights and interests of citizens and legal persons'[38] in China. Modeled on the German Civil Code, the GPCL was one of the most market-centric pieces of legislation from the decade as it 'contemplate[d] a universe of equal actors forming and altering legal relationships by acts of free will.'[39] Among other things, the GPCL formally recognized both natural and non-state enterprises as having the capacity to contract,[40] and established basic agency principles, including the ability for a principal to enter contracts through an agent.[41]

b. Business association laws The liberalization of China's economy during the 1980s also necessitated that China define what types of entities could lawfully participate in economic transactions in China.[42] At the outset of the decade, there was no meaningful 'business organization' system in China.[43] China's 1982 Constitution provided for three ways to organize a business: (1) socialist public ownership;[44] (2) collective ownership;[45] and (3) 'the individual economy of urban and rural working people, operated within the limits prescribed by law.'[46] This third category was expressly characterized by the 1982 Constitutions as a 'complement to the socialist economy.'[47] There were no laws or policies at that time for organizing larger private business organizations to operate in this 'individual economy.'[48] As a result, many private businesses operated as 'red hat' enterprises – which were private businesses 'formally registered as collective instead of private in exchange for a fee paid to the local government for protection against predation and to qualify for various benefits available only to the public sector.'[49] In 1986, the GPCL helped to advance the business organization issue by recognizing that private forms of Chinese 'legal persons' could exist[50] – laws permitting the existence of foreign-invested enterprises had already been on the books since 1979.[51] The 1988 Amendments to the Constitution, coupled with various regulations implemented in 1988, completed the legalization of some rudimentary form of Chinese private enterprises – many of which had been operating for years outside the law.[52]

1.2.3   The law's transition to a traditional tool for economic development was gradual

China's transition to a market economy was radically different from the rapid market transition, frequently referred to as 'shock therapy,' that took place in Russia and Eastern Europe in the early 1990s. The basic idea behind shock therapy was quite simple. Following an initial stabilization period, the former planned economy was transformed into a market-based economy almost instantaneously, rather than in a more gradual fashion. Government controls over the economy (e.g., price controls, subsidies to state-owned enterprises, and employment controls) were quickly eliminated, while state-owned assets were rapidly privatized.[53] The old mechanisms for governing the countries' economic actors were rapidly dismantled with the expectation that new market-based mechanisms (e.g., property rights and contract enforcement) would rapidly develop to fill the void. Many blame the shock therapy strategy for Russia's severe post-Soviet economic troubles.

With China, the much slower transition to a market economy meant that China's old governance mechanisms were largely maintained at the outset of the reform era.[54] The old mechanisms continued to operate because they were needed to govern the state sector, which remained the dominant force in China's economy. Rather than undergo a rapid elimination, the old mechanisms were allowed to 'whither gradually over the following decades.'[55] One of the more distinctive characteristics of China's early reform movement was the tendency to employ a dual-track system to various segments of the economy. Economic actors that were subject to China's compulsory planning system were still subject to their assigned output quotas. If they had the ability to generate 'above-plan' output, however, they were able to sell the excess at market prices.[56] In addition to increasing output, the dual-track system introduced most of China's state-run factories and enterprises to the market and allowed them a transition period to adapt to market processes.[57] The dual-track system also created a mechanism for state sector and non-state sector entities to enter into cooperative transactions.[58]

In this environment, the secure expectations that the law can provide was only truly necessary at the margin of economic activities, which greatly reduced the incentive for developing a strong, formal legal system. Existing bureaucratic mechanisms for enforcing transactions, even if highly imperfect and wasteful, were capable of managing the majority of economic transactions. While there were numerous benefits to China's gradual, step-by-step approach to introducing market reforms and developing the supporting economic legal system, one of the practical results was a relatively underdeveloped economic legal system in the early 1990s when China launched its more concerted market-based reforms. This is not meant to be a criticism of China's legal system in the early 1990s, which demonstrated incredible progress since the commencement of the reform era, but it does help to place in context the relative starting point of China's legal system at the time when market-based reform really took off.

2. THE CURRENT REFORM ERA AND THE DEVELOPMENT OF A MODERN LEGAL SYSTEM (1992 TO PRESENT)

2.1 The Market Reform Movement Prevails

China's current stage of economic reforms is frequently traced to Deng Xiaoping's 'Southern Tour' of three of China's original special economic zones (Shenzhen, Zhuahai and Shantou) in early 1992. China experienced a period of 'conservative ascendency' from 1989 through 1991 following the Tiananmen Square crisis.[59] The conservatives were unable to maintain that momentum and Deng's visits to the special economic zones, which highlighted the success of China's market-based economic reforms, helped 'restore the government's commitment to economic reform and tip the balance of political power in Beijing.'[60] In October of 1992, the 14th Congress of the Communist Party met and officially endorsed a 'socialist market economy' as the goal for China's reform movement. That gesture was followed in 1993 by the Second Amendment to China's 1988 Constitution which, among other things, amended Article 15 to replace the reference to a 'planned economy' and a 'supplementary role' for the market with a simple and clear-cut declaration that China practices a 'socialist market economy' (see Table 4.1). The 1993 Amendment also eliminated the duty to fulfill the State plan that was formerly set forth in Articles 16 and 17 of the 1988 Constitution (see Table 4.1). While the State plan requirement in Articles 16 and 17 was not 'legally binding' as one would normally think of that concept, the elimination of the references was symbolically important.[61]

Since 1993, the central role of markets in China's economy has not been seriously challenged. The dual-track system was quickly eliminated and China's leadership has worked to 'create an economic legal system that fully and explicitly embraces the private sector as an important component of the economy.'[62]

Table 4.1   1993 amendment to Articles 15, 16 and 17 of China's 1988 Constitution

Prior Text	Amended Text
Article 15	
The State practices economic planning on the basis of socialist public ownership. It ensures the proportionate and coordinated growth of the national economy through overall balancing by economic planning and the supplementary role of the market. Disturbance of the socio-economic order or disruption of the State economic plan by any organization or individual is prohibited.	The State has put into practice a socialist market economy. The State strengthens economic legislation, improves macro-regulation and control, and prohibits according to law any organization or individuals from disturbing the social economic order.
Article 16	
State enterprises have decision-making power with regard to operation and management within the limits prescribed by law, on condition that they submit to unified leadership by the State and fulfill all their obligations under the State plan. State enterprises practice democratic management through congresses of workers and staff and in other ways in accordance with law.	State-owned enterprises have decision-making power with regard to their operation within the limits prescribed by law. State-owned enterprises practice democratic management through congresses of workers and staff and in other ways in accordance with law.
Article 17	
Collective economic organizations have decision-making power in conducting independent economic activities, on condition that they accept the guidance of the State plan and abide by the relevant laws. Collective economic organizations practice democratic management in accordance with law. The entire body of their workers elects or removes their managerial personnel and decides on major issues concerning operation and management.	Collective economic organizations have decision-making power in conducting independent economic activities, on condition that they abide by the relevant laws. Collective economic organizations practice democratic management and in accordance with law, elect or remove their managerial personnel and decide on major issues concerning operation and management.

Note: General source for English translation: Asian Legal Information Institute, Laws of the People's Republic of China, Amendment to the Constitution of the People's Republic of China 1993, http://www.asianlii.org/cn/legis/cen/laws/attcotproc1993567/.

2.2 The Law Takes a Prominent Role in Reform

Whereas the law played a somewhat reduced role in the early reform period, the law began to catch up with economic activities in the current reform era. The law is no longer limited to legitimizing economic activities that have already been developed. The law in China now has the more significant capability to improve the efficiency of existing economic activities and even create entirely new activities. Developing the institutions of a market economy, including a properly functioning legal system, has been a central focus of Chinese policymakers during the current reform era.[63] The law is no longer an afterthought, but instead is a central focus of the reform effort.

Much of the current reform era has focused on empowering private (non-state-owned) companies to play a critical role in the economy (rather than being relegated to a 'supplementary role') and encouraging state-owned enterprises to operate more like private sector companies. In such an environment, the law becomes considerably more important as a governance tool. The old, bureaucratic governance mechanisms were no longer effective in this new environment. There were too many economic actors pursuing too many potential transactions for the old mechanisms to govern the system and provide the secure expectations needed to properly motivate the various private and non-private companies and the individuals running those companies. Unlike in the early reform era when a well-functioning legal system was not entirely necessary, the increasingly complex economic relationships of the current reform era necessitated a consistent, rules-based system that applied equally – or, at least, somewhat equally – to all economic actors.[64]

During the current reform era, China has made many important changes. First, China has succeeded in adopting a set of modern civil codes that cover a vast array of economic activities. Second, China has succeeded in training hundreds of thousands of lawyers. And third, China has dramatically improved the competency of its court system (including by professionalizing the judiciary).

2.2.1 Adopting modern civil codes

During the current reform era, China has adopted a massive number of the laws that are typically associated with a market economy (see Box 4.1 for an overview of China's formal rule-making structure). Consider just some of the major legislation adopted by China since 1992:

- *Business association legislation:* China has adopted a series of laws that now allow for a variety of modern forms of business

BOX 4.1   CHINA'S FORMAL RULE MAKING STRUCTURE

China's formal rule making structure is set forth in the Constitution, the Law on Legislation (2000) and the State Council's Regulations on the Procedure for the Enactment of Administrative Regulations (2001). Laws are passed by the National People's Congress (NPC), which is China's elected parliament. Members of the NPC are elected through an indirect election process. Local elections are held to choose representatives who then elect the members of the NPC.[1]

The NPC has long been criticized by Western commentators as being little more than a 'rubber stamp' organization. Laws do not tend to originate from the NPC, but instead are initiated and drafted by various governmental bodies that present the legislation for approval.[2] While it is unusual for the NPC to reject legislation submitted for its approval,[3] the 'rubber stamp' criticism is unfair. The power of the NPC appears to have increased through the 1990s and members of the NPC are now more prone to exercise their own judgement when deciding on legislation rather than simply approving laws submitted by the Party.[4]

There are a 'bewildering array of bodies that have the right or the practical power to make rules of varying degrees of binding effect'[5] in China. The ministries of the State Council, other bodies of the central government and the local government all have the ability to pass regulations. In theory, statutes passed by the NPC are the most authoritative rules in China other than the Constitution. In practice, however, such superiority of the law does not always win out.[6]

Notes

1. Gregory C. Chow, China's Economic Transformation, 2nd Edition 369 (2007).
2. OECD (2003), *Investment Policy Reviews: China – Progress and Reform Challenges*, p.113.
3. *Id.*
4. Chow, *supra* note 1, at 369.
5. Donald Clarke, Peter Murrell and Susan Whiting (2008), 'The Role of Law in China's Economic Development', in Loren Brandt and Thomas G. Rawski (eds), *China's Great Economic Transformation*, p.394.
6. *Id.*, at 394–395 and 399–400.

organizations for private sector companies and provide a reasonable 'corporate governance' system. Such legislation includes the Company Law (1993, substantially amended in 2005), the Partnership Enterprise Law (1997, substantially amended in 2006), and the Sole Proprietorship Enterprise Law (1999).

- *Contract Law:* The Contract Law (1999), which dramatically modernized China's system of contract law and unified China's three prior, major contract laws – Law of Economic Contracts (1981), Law of Foreign-Related Economic Contracts (1985) and the Law of Technology Contracts (1987) – into a single contract law.
- *Property Law:* The Law of Property (2007), which gives private property rights equal status to state and collective property rights.[65]
- *Banking legislation:* The Law on Commercial Banks (1995, substantially amended in 2003).
- *Bankruptcy legislation:* The Enterprise Bankruptcy Law (2006).
- *Capital markets legislation:* The Securities Law (1998, substantially amended in 2005).
- *Competition legislation:* The Law against Improper Competition (1993) and the Anti-Monopoly Law (2007).
- *Consumer protection legislation:* The Product Quality Law (1993, substantially amended in 2000), the Law on the Protection of Consumer Rights and Interests (1993) and the Law on Advertisements (1994).
- *Debtor/creditor legislation:* The Guarantee Law (1995) and the Law of Negotiable Instruments (1995, substantially amended in 2004).
- *Foreign trade and foreign investment legislation:* The Law on Foreign Trade (1994, substantially amended in 2004).
- *Insurance legislation:* The Insurance Law (1995, substantially amended in 2002).
- *Intellectual property legislation:* The Patent Law (substantially amended in 1992, 2001 and 2008), the Trademark Law (substantially amended in 1993 and 2001) and the Copyright Law (1990, substantially amended in 2001).

To top things off, China's Constitution was amended in 1999 to declare that China will be governed 'according to law' and that China is a 'a socialist country ruled by law.'[66] While this rule of law provision of China's Constitution was probably best categorized as aspirational in 1999, it is highly indicative of the increased prominence of the law in the current phase of the reform era.

It is hard to overstate the impressiveness of China's legislative accomplishments since the commencement of the current reform era. In less than two decades, China has adopted a bewildering array of modern, sophisticated legislation to support its new, market-based economy. We are not aware of any country in history that has been able to implement such an impressive amount of high-quality legislation in such a short time frame. Adopting laws, however, is not the same thing as developing a strong legal system. Legislation alone does not change economic behavior.[67]

a.   Enforcement and generally accepted norms of behavior   For a law to change economic behavior, it needs to be followed. Ideally, laws are aligned with the relevant community's generally accepted norms of behavior (GANB), because it makes enforcement of the law much simpler and less expensive. In those cases, economic actors follow the rules primarily because they respect the rules, not because formal enforcement mechanisms compel them. Consider the example of home security for most homeowners in the United States. Throughout the United States, there are formal legal rules that clearly make it illegal to trespass in someone else's home and steal property from that home. Formal mechanisms play only a minor role, however, in the enforcement of the homeowner's rights. Relatively speaking, why are there so few home break-ins in the United States? Is it the deterrent effect from potential criminal prosecution? Probably not, since most home burglaries go unprosecuted. Is it the quality of the home security system? For the vast majority of homes, the home security system consists of closing and locking your doors and windows. Such a security system is really nothing more than a minor nuisance to a motivated burglar, since most windows can be easily broken. The primary enforcement mechanism for protecting the sanctity of the home is the presence of informal, behavioral norms. Most people respect the sanctity of the home and would feel bad if they were to break into someone else's home and steal their property. Shame prevents people from acting as burglars much more than the repercussions from a formal judicial system. Because social norms play such a large role in dictating human behavior and dissuading burglary, local governments do not need to dedicate that many resources to preventing burglary. For example, police are not stationed on every corner of every neighborhood, which keeps the overall cost of enforcement relatively low.

On the flip side, when the formal institutions are not well aligned with the relevant community's GANB, enforcement can be much more difficult and expensive because the economic actors do not naturally respect the rules. A classic example of this phenomenon is illegal downloading of digital music. Such downloading clearly violates basic principles of copyright law and constitutes stealing of someone else's property. When you ask most illegal downloaders if such downloading is wrong, however, the answer is almost universally 'no.' They do not 'think' they are stealing. The legal rule is not aligned with the relevant community's GANB, as the relevant community has little respect for the ownership rights of the copyright holders. The result is that the illegal downloading of digital music has been rampant, relatively speaking, over the last decade. When legal rules are not aligned with community behavioral norms, policymakers tend to take three basic approaches:

- *View the legal rules as an aspiration:* policymakers may simply accept a reduced level of compliance for a period of time, while hoping that improved compliance/cultural acceptance will develop over time. Where the rules are needed in the short term to incentivize important behavior, however, such a passive approach might not be sufficient.
- *Educate:* policymakers can try to educate the relevant community on the reason for the legal rules and thereby try to align their behavioral norms with the legal rules. In China, for example, the government has spent a fair amount of resources to try and educate the general population on the importance of IP and respecting the rights of IP holders.
- *Commit more resources to compliance:* finally, policymakers can try to force compliance by committing substantial resources to the compliance effort.

Because law is a relatively new tool for motivating economic behavior in China, there is a much higher likelihood that a significant percentage of economic actors will have a low regard for the law and seek to avoid compliance. Chinese policymakers seem to understand this problem, as most of their major legal initiatives appear to knowingly follow one, or more, of the three above approaches.

b. Individual laws do not operate in a vacuum To make matters more complex, individual laws do not operate in a vacuum, but instead operate within an overall institutional environment. The success of any one law will likely be highly dependent on the strength, or weakness, of numerous other complementary laws and institutions. Even if the country's property law system is extremely well designed, for example, it is not likely to generate anything close to a proper level of productive exchanges without a correspondingly well-designed judicial system to enforce the property rights; a contract law institution to facilitate the exchanges; and strong financial institutions to fund both property development and property exchanges. Similarly, financial institutions are unlikely to be effective without a well-designed property institution, judicial system, contract law institution and bankruptcy institution, as well as a sound system of financial market regulation.

This interdependence can prove quite challenging for policymakers – in particular for countries that are just beginning to develop meaningful market-based institutions. In many countries, policymakers may not have the luxury of slowly layering on the complementary laws and institutions over an extended period of time. If a country wants to build market-based financial institutions, for example, it needs an extensive array of other institutions to give the financial institutions a reasonable opportunity to function properly. That places policymakers in a difficult dilemma. On the one hand, they can try to implement a full array of institutions at one time. That approach has a number of obvious problems. Designing effective institutions is very difficult, even when policymakers are developing a very limited set of institutions. When trying to develop a full array of institutions, it is unrealistic to think that policymakers will be able to develop those institutions with any level of true care – which makes it less likely that the institutions will be properly followed and enforced. On the other hand, policymakers can take a slower, more gradual approach to institution building. The problem with that approach is that well-designed institutions may initially prove to be unsuccessful due to the lack of complementary institutions. If this second approach is taken, policymakers need to be patient with the process and not draw wrong conclusions from initially poor results. China's policymakers have clearly leaned towards the latter, slower approach to building its legal system and have wisely demonstrated the necessary patience.

2.2.2 Training lawyers

One of the biggest changes in China's legal system is the blossoming status of lawyers in China. While lawyer jokes are popular (and sometimes even funny), it goes without saying that a sophisticated, modern legal system cannot operate without well-trained, highly-skilled lawyers. During the early reform period, China did not have nearly enough lawyers. In 1985, seven years into the reform era, China had just over 13 000 qualified lawyers in the entire country – and roughly half of them were part-time lawyers.[68] By the end of 2006, that number had multiplied tenfold to 130 000 lawyers working in 13 000 law firms.[69] China's strong commitment to legal education has helped fuel this increase. China has established a comprehensive legal education system that provides bachelor's (LLB), master's (LLM) and doctoral (PhD) degrees in the law. As of 2006, China had 603 universities and colleges offering bachelor of law degrees and 300 000 undergraduate law students.[70] In addition, 333 Chinese institutions offer master's law degrees and 29 offer doctoral law degrees.[71] China is well on the way to populating its legal profession with a substantial number of lawyers to perform the various legal services that are needed for a legal system to be considered functioning. There is no reason to doubt that this increase has been driven primarily by demand for legal services.[72]

As we have seen elsewhere in this book, there is frequently a concern in China that 'quality' may be sacrificed in order to achieve substantial 'quantity' increases. There are a lot of lawyers in China now, with many more on the way – but are they any good? There are no statistical data available to answer that question definitively. We are left with loads of anecdotal stories about various parties' experience with Chinese lawyers – with some stories being positive and some being negative. Overall, based particularly on Hong Shen's experience as a lawyer and law professor in China since 1983, we believe those varied stories paint a relatively accurate picture of the quality of lawyers of China. Just like in any of the developed countries, there are good Chinese lawyers and bad Chinese lawyers, and we see little reason to believe that the ratio of bad lawyers to good lawyers is somehow worse in China than it is in the developed world. The quality of lawyers produced by China's law schools is generally good. The competition to get into law school and the competition to become a law professor in China is fierce, which generally leads to quality incoming students, quality instruction and, not surprisingly, quality graduating law students.

Even if China's law schools were not producing truly high-quality lawyers – although we think they are – there are other reasons to take comfort in the overall quality of China's lawyers. As more lawyers operate in China, the more the law gets enforced, tested and expanded.[73] As these new law school graduates become lawyers in China, they are going to be looking for things to do. They will be looking for legal rights to enforce and for deals to structure, and they will begin to push the edges of what is legally permissible – which will create demand for new and better laws and more legal services.

Much of the art of good 'business lawyering' comes from accumulating valuable experiences about the law and how the law can be valuably applied to various business settings. When more valuable business lawyering experiences are made available to a population of lawyers, the overall quality of that group of lawyers should correspondingly improve. In China, the last decade has generated an almost perfect storm of inputs for creating a very high level of valuable business lawyering experiences that we believe has profoundly and positively impacted the quality of Chinese lawyers. First, the number of practicing lawyers in China has reached a critical mass so that enterprises and organizations can now integrate lawyers and legal services into their normal activities. Second, the quantity and quality of Chinese laws and regulations have now reached a level

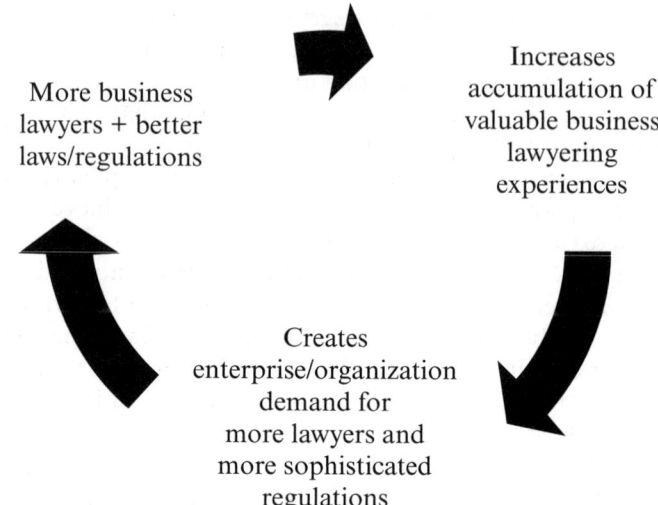

More business
lawyers + better
laws/regulations

Increases
accumulation of
valuable business
lawyering
experiences

Creates
enterprise/organization
demand for
more lawyers and
more sophisticated
regulations

*Figure 4.1   Virtuous cycle that is improving quality of Chinese business
lawyers*

where business lawyers can rely on, and creatively use, the law to create
value in a wide variety of settings, which further motivates enterprises and
organizations to ramp up their usage of lawyers. Finally, as lawyers accumulate these experiences and become more proficient, it creates a virtuous
cycle (see Figure 4.1). An increase in better lawyers creates enterprise/
organization demand for more lawyers and more sophisticated regulations, which increases the ability to accumulate additional valuable business lawyering experiences.

2.2.3   Professionalizing the judiciary

The judiciary plays a critical role in market-based economic systems by
enforcing the legal rights that are central to market exchanges. Like those
of many European countries, such as France and Germany, China's trial
court system is structured as an 'inquisitorial' rather than an 'adversarial'
system. Judges play a much more active role in inquisitorial systems. In an
adversarial system, the resolution of disputes relies heavily on the advo-
cacy abilities of the litigants, which in a civil case are the plaintiff and the
defendant. The plaintiff and defendant plead their cases, and the resolution of the dispute is left to 'impartial referees.' An impartial jury will typically determine the facts of the dispute and an impartial judge will make
the determinations of law. In an inquisitorial system, the judge (or more
specifically the court system) takes a much more active role in investigating

and determining the facts of the case. Judges will question witnesses and suspects, order searches and direct investigation paths. In the end, it is the inquisitorial judge that makes the findings of fact. The greater responsibility for judges in an inquisitorial system increases the importance of both their competence and their independence from undue influences.

China's judiciary has frequently drawn fire from critics that it lacks the competence and independence needed to play a proper role in China's legal system. Albert Chen, a law professor at the University of Hong Kong, explains:

> [A]s in the case of many developing countries, there have existed serious concerns regarding the professional quality and judicial ethics of the Chinese judiciary, corruption, susceptibility to political interference (which in the case of China may sometimes be legitimized by the supreme principle of the leadership of the Party in all spheres of state activities, including the judicial function) and problems in the enforcement of judgments.[74]

The competency/independence problem within the Chinese judiciary is largely historical. When China's court system was reconstituted in 1979, there were nowhere near enough competent individuals to serve as judges. As we explained earlier in this chapter, the Cultural Revolution wreaked havoc on China's court system and its law schools, which severely disrupted the normal talent pools for experienced and well-trained judges. The government had to recruit its judges from a variety of non-traditional sources, such as demobilized military officers, school teachers and administrative officers.[75]

There is little debate that China's judiciary had significant issues at the outset of the reform era. Since that time, however, China has taken numerous concrete steps to improve the competency/independence of of its judiciary and to create a career-judge system that is similar to the European model. During the 1980s, China established national senior judge training centers. In 1995, the NPC passed the Law on Judges 'to enhance the quality of judges, to strengthen the administration of judges, and to ensure that the People's Courts independently exercise judicial authority according to law, that judges perform their functions and duties according to law and that law is administered impartially.'[76] Among other things, the Law on Judges established minimum standards for all new judges. In 2000, a competitive process for selecting judges was implemented, which was quickly followed by a significant revision to the Law on Judges in 2001 that required new judges to meet specific legal education and experience requirements[77] and to pass a state uniform judicial examination.[78] The first judicial examination was administered in March 2002.[79] The Supreme People's Court is also working to improve the performance of China's judiciary and has implemented a number of reforms, including the adoption of a collegiate court system that is aimed at increasing independence and impartiality.[80] The Supreme People's Court has also issued a number of five-year reform outlines that establish a variety of reform tasks that are to be pursued. The latest such reform outline was issued in March 2009 and covers the 2009–2013 period.

In spite of these efforts to professionalize the judiciary so that it may play an effective role in China's legal system, many critics continue to attack the quality and independence of China's judges. As with the issue of lawyer quality in China, there is no perfect tool for determining judicial quality/independence, which allows for everyone to have an opinion on the issue and express that opinion very authoritatively – even if the opinion is woefully inaccurate. To make matters worse, China is a very large country with a large number of judges operating in diverse settings and regions throughout the country, which provides a rich reservoir of potential anecdotal stories to support opinions, particularly negative opinions. For critics of China's judiciary, one need only find a handful of corrupt or incompetent judges, and then project that corruption/incompetence throughout the system. Since corruption and incompetence have historically been problems with China's judiciary, such criticisms tend to find an uncritical audience that is very receptive to such charges.

So, where does that leave the quality of China's judiciary and its ability to enforce the legal rights that are central to a market-based economic system? We do not profess to have the ultimate answer to that question, but our belief – which we are not making as an authoritative proclamation, just a belief based on Hong Shen's 20+ years of experience operating in the Chinese legal system and our accumulation of mountains of positive anecdotal stories – is that China's judiciary operates at a sufficiently competent and independent level to support a market-based economic system. Is China's judiciary perfect? Of course not. As with every judiciary,[81] there are bad judges, but those judges are far outweighed by competent, thoughtful judges who are sincerely working to properly apply and enforce the law over the wide variety of cases that come before them.

3.   ASSESSMENT OF CHINA'S CURRENT LEGAL SYSTEM AND ITS ABILITY TO SUPPORT MARKET TRANSACTIONS

Market-based strategies, such as a Bayh–Dole strategy, are based on voluntary exchanges between informed parties. While this is a bit of an oversimplification, the most minimal institutional requirements for a healthy market environment are well-defined, secure property rights and enforceable contracts.

- *Well-defined, secure property rights*: well-defined property rights that are easily enforceable (including against the government) are the most fundamental element for a successful market-based system. They establish the parameters of 'what' is being exchanged.
- *Enforceable contracts*: a market-based system also requires consistently enforceable agreements that can be relied on to design and execute the actual property exchanges.

In developed countries, the formal legal system plays a central role in creating an optimal property and contract regime that facilitates productive economic exchanges. This section will examine whether China's legal system has sufficiently evolved to provide the level of property/contract protection that is needed to support market transactions generally. Chapters 5 and 6 will then consider China's IP regime and its specific Bayh–Dole legal structure.

3.1 Secure Property Rights

The security of property rights presents a bit of a paradox. For much of the reform period, China's legal system by itself did not provide the security of property rights that one normally associates with a well-developed market economy,[82] and yet the economy, and market-based elements of the economy, grew spectacularly.

The quality of China's early private property legislation was generally very weak. Over time, however, modern property legislation has been developed, with intellectual property legislation being some of the earliest private property ownership legislation in China. China's 1984 Patent Law, for example, was truly at the forefront of China's general evolution during the reform era towards a private property system.[83] In 1984, the Patent Law forced China's policymakers to wrestle with many of the issues that they have had to grapple with in expanding private property rights generally in China. Policymakers had to find comfort that patents were not incompatible with China's socialist economy, and they had to overcome their fears that foreigners would end up owning the most valuable patents.[84] Throughout the 1990s and 2000s, China's property legislation substantially expanded the scope and security of private property ownership. For example, China's entire IP legislation system was massively upgraded beginning roughly in 1990[85] and has resulted in China now possessing a thoroughly modern set of IP laws and regulations. On the tangible property rights side, the move towards greater, and more secure, private property culminated in 2007 with the enactment of the much-awaited Property Law of the PRC. As Professor Mo Zhang explains:

> The passage of the Property Law of China on March 16, 2007 marked an historic change in the country from public to private with respect to property rights. Effective on October 1, 2007, the Property Law for the first time in Chinese history grants an equal protection to both public and private properties, breaking up the orthodox ideology in favor of public ownership against private ownership and individual liberty.[86]

While China's current private property legislation is by no means perfect – and we are not aware of any country whose private property legislation is perfect – it does respect the notion of private property rights and provide property owners, including non-government property owners, with a reasonable ability to own and improve property and collect the rewards from their investment in that property. As with all countries, questions remain about poorly drafted aspects of China's various property laws and regulations,[87] the impact of overlapping (or conflicting) rules that are sometimes adopted by various governmental bodies,[88] and the ease with which the government can exercise its taking powers.[89] Overall, however, China's property legislation appears to be evolving into what one expects for a market-based economy.

3.2 Enforceable Contracts

Reduced to its simplest concept, contract law deals with legally enforceable promises. From a purely legislative standpoint, the unified Contract Law (1999) provides a reasonably well-written contract law that is suitable as a modern contract law system. The unified Contract Law's approach to contract formation,[90] the validity of contracts,[91] performance and breach of contract,[92] and remedies[93] are all reasonable. For a contract system to be truly economically useful, however, contracting parties need to be able to rely on both the contract law and the agreements that they enter into as a result of that contact law. Contracts, and contract rights, must be consistently enforceable.

3.2.1 Guanxi
There is a common perception that legal enforcement of contracts may not be all that important in China. Instead of relying on the formal legal system to create and enforce business agreements, some Chinese analysts have suggested that Chinese business agreements are much more likely to depend on informal social ties between the parties, or *guanxi*.

The Chinese word '*guanxi*' . . . refers to the concept of drawing on connections in order to secure favors in personal relations. It forms an intricate, pervasive relational network which the Chinese cultivate energetically, subtly, and imaginatively. It contains implicit mutual obligations, assurances, and understanding, and governs Chinese attitudes towards long-term social and business relationships. Broadly, *guanxi* means interpersonal linkages with the implication of continued exchange of favors. *Guanxi* is therefore more than a friendship or simple interpersonal relationship, it includes reciprocal obligations to respond to requests for assistance.[94]

Informal social networks are employed in all business cultures, with the U.S. old boys' network, Japanese *wa* and Korean *inhwa* serving as classic examples.[95] Based on the trust and cooperation that such networks facilitate, they can serve a number of valuable functions to encourage economic transactions. They can help to facilitate the gathering and dissemination of valuable information amongst members of the network. Such networks can also provide an extra-legal mechanism to encourage trustworthy behavior between members of the network, which creates valuable opportunities to use members of the network as trusted agents for economic transactions.[96] On the flip side, social networks can also be value-destructive to an economy by reducing competition – e.g., by discriminating against competent actors outside the network – and increasing transactions costs when the costs involved with joining and operating the social network are greater than the costs of providing 'network-type' services through more formal mechanisms such as entering into legal contracts.

3.2.2   Written contracts, rather than *guanxi*, are the norm in China for business transactions

For countries with weaker formal institutions, reliance on informal social networks to facilitate economic transaction is not uncommon.[97] As countries develop, and the range of valuable economic activities and partners increases, informal institutions such as social networks tend to decrease in value and are edged out by more formal institutions – such as a stronger legal system.[98] China has followed this typical development path. While *guanxi* persists in social matters, the use of formal, written contracts has become the norm for business transactions in China.[99] Hong Shen explains:

Since at least the late 1990s, the limitations of *guanxi* for business transactions have been quite evident in China. Business transactions in China are conducted through formal, written agreements just as one would expect in the United States or any other developed country. Can strong personal relationships help parties to enter into business transactions? Of course, but that is the case in any country. Trusting your counterparty makes it easier to enter into a deal. It is important to stress, however, that even when *guanxi* helps parties to make a deal, a well-written contact is still crucial. I have litigated a number of contract disputes where the parties entered into the agreement based on a *guanxi* relationship, and then signed a sloppy, written agreement. As is frequently the case, misunderstandings developed in the relationship, and the written contract was not very useful in resolving the misunderstanding. There really is never a reason for not entering into a well-drafted contract that clearly expresses the parties' intentions. In my experience, the importance of well-drafted, written contracts is well understood in China.

3.2.3   Appreciating the shift to formal mechanisms

Few would disagree with any of the following assertions:

- Property and contract rights in China are better defined and more secure today than they were 10 or 20 years ago;
- The quality of China's property and contract legislation is better today than it was 10 or 20 years ago; and
- The competency of Chinese courts is better today than it was 10 or 20 years ago.

Yet, many foreign analysts are not yet ready to attribute the improved definition/security of Chinese property rights to China's improving legal system. Even analysts that are generally favorable towards China's evolving property system tend to make crystal clear that they hold serious reservations about the formal legal system's role in that improvement. It is as if the improvements in property legislation and the courts are mere 'gestures' and not fundamental reforms that are greatly increasing the rule of law in China. We are not entirely clear why it is so difficult to applaud the improvement China's formal legal system has achieved with respect to property rights. Is it a lack of trust in the motives of the CCP or the central government? Is it simply a matter of habit – analysts have been critical of China's legal system for so long that it is difficult for them to alter their fundamental position? Moreover, there is always the convenient explanation that informal mechanisms, rather than China's legal system, provide the confidence needed for business transactions in China and not the law.

Fundamentally, we believe this skepticism about the improving legal system's impact on property and contract rights in China stems from a deeper skepticism about the rule of law in China. Many Chinese analysts continue to under-appreciate how much progress China has made in developing the rule of law for its economic affairs. Short of providing skeptics the ability to operate in China's business law over an extended period of time, it is difficult to definitively prove the meaningful progress that China has made over the last few decades in making the rule of law an integral part of economic transactions. Dan Harris, a U.S. business lawyer with extensive Chinese experience and the author of the popular 'China Law Blog,' recently posted about this issue. Dan summarized a conversation with one of his international clients who has been doing business in China for roughly five years:

> [My client was telling me] how an American competitor of his had gotten into legal troubles and was on the verge of pulling out of China. My client told me he thought his competitor had brought the problems onto his company by believing he could get away with not following Chinese laws. We then talked about how when it comes to China's laws relating to business, they are actually usually fairly clear and actually usually not all that bad. We both agreed that companies that follow China's business laws overwhelmingly avoid problems. But, we both also agreed that what we were discussing had little to nothing to do with China's non-business laws and little or nothing to do with the corruption. In other words, China's business laws, as written are good and following them usually insulates you from problems. But, not following them and having the right connections (which damn few foreigners have, despite their thinking otherwise) can oftentimes serve to avoid problems also.[100]

We believe that Dan captures the situation in China very nicely.

4. DESPITE THE UNFAIR CRITICISM, CHINA'S LEGAL SYSTEM IS MORE THAN SUFFICIENT TO SUPPORT THE MARKET-BASED TRANSACTIONS OF A BAYH–DOLE SYSTEM

It is difficult to overstate how much progress China has made in developing a market-supporting legal system. In 1992, when China's serious legal reform effort truly started, the legal system's role in China's economy was weak. Less than 20 years later, China has a reasonably-functioning, modern legal system that supports the world's third-largest economy. It is by no means a perfect legal system, but the law is now an integral part of China's economic affairs. In spite of the considerable progress that has been made with China's legal system, Western sentiments about it continue to be exceedingly and inappropriately negative.[101] The least informed still question whether China even has a 'real' legal system. While China's legal system has its flaws and remains a work-in-progress, the criticisms it tends to receive are frequently out of proportion to the reality.

The Western media, which are the loudest critics of China's legal system, tend to focus almost entirely on human rights violations, the plight of dissidents[102] and the imprisonment of reporters. It goes without saying that each of these issues is extremely important and merits news coverage. Yet, such coverage paints a picture for many that China's entire legal system is weak and ineffectual, which is simply not true. Few would try to judge the entire American legal system based on its worst judicial actions (e.g., judicial support for Jim Crow laws in the southern states of the United States or judicial support for Japanese internment camps during World War II), and yet that is what seems to occur when China's legal system is analyzed. Focusing solely on the Chinese legal system's struggles does not do justice to the progress that has been made over the last two decades.

The foreign business community is another frequent critic of China's legal system. Chinese legal scholar and former expatriate lawyer in China, Randall Peerenboom, helps to place in context many of the foreign business community's complaints:

> While [foreign business executives and their trusty sidekicks, expatriate lawyers,] will usually acknowledge that China's legal system has made considerable progress in the last twenty years, the dominant theme for many is that the system remains so riddled with problems that it is questionable whether it makes sense to even speak of the system in terms of rule of law. Furthermore, business people and lawyers are likely to turn to the media to complain when the system fails to function as they expect it to, or at least hope it would. Conversely, when all goes smoothly, they are likely to take it for granted. Furthermore, the views of lawyers are likely to reflect their own experiences. The billing rate of lawyers working in major international firms is very high. Given this fact, companies are not likely to seek the advice of outside counsel except on cutting edge projects or complicated issues where the law is unclear or there are other obstacles involved. Thus, lawyers are likely to encounter on a daily basis the tough cases rather than the easy ones.[103]

A simpler explanation of the foreign business community's criticisms may be their lack of effort to thoroughly learn China's business laws before entering into business transactions in China. Dan Harris provides another example that illustrates how easy it is to make false claims that China is a 'country without laws,' and how such false claims are willingly spread by the media:

> Three or four years ago, . . . a whole spate of articles came out on foreign companies whose trademarks had been 'stolen' by Chinese companies. I realized that most of these articles were leaving out an essential fact in that they did not say whether the foreign company had actually registered its trademark in China or not. So I called some of the reporters who had written these articles and every single time they did not know whether the companies had registered their trademarks in China or not. Seeing as how one cannot have one's trademark

stolen without first owning such a trademark and owning a trademark in China requires registering the trademark in China, these articles had it all wrong. Their thrust should not have been, 'China is a country without laws.' Their thrust should have been on how foreigners were going into China without first making any real effort to understand China's business laws.[104]

On the academic front, foreign legal scholars appear to be recognizing, more and more, the progress being made with China's legal system. Interpreting foreign legal scholarship about China's legal system, however, can be difficult for non-legal scholars. To begin with, legal scholars tend to spend their energy identifying problems with legal systems, rather than applauding strengths. So, scholars who may consider themselves to be highly complimentary of China's progress in developing its legal system may sound very negative to non-legal scholars since the focus of their research will be on what should be corrected, rather than on all that is right with the system. To compound matters, there is a tendency of foreign legal scholars to measure China's legal system against 'an idealized version of liberal democratic rule of law that does not exist in reality anywhere.'[105]

Overall, we believe that descriptions of China's business law system require more balance. We also urge a greater appreciation of the progress that has been made in China's legal system. It is simply inaccurate for China commentators to start from the premise that China's legal system is terribly flawed and/or non-functioning. Starting from such a negative position encourages foreign businesspersons to underestimate the importance of law in China, and thereby fail to properly protect their legal interests when operating in China. It also encourages journalism and academic research that tends to repeat criticisms of China's legal system without sufficient investigation. Positive statements about China's legal system need strong proof, but criticisms (i.e., China's legal system does not really support property rights) can be repeated with little back-up. A more balanced approach to analyzing China's legal system allows for flaws in the system to be more clearly identified, without the distraction of the overly negative hyperbole, and for China to continue its incremental, steady progress toward a rule of law-based approach to regulating its growing market economy.

While China's legal system is by no means perfect, it is sufficiently supportive of market transactions for a Bayh–Dole system to develop and flourish. Over the last 20 years, China has developed an impressive number of sophisticated laws and regulations to support market transactions – including business associations legislation, contract law, property law, banking legislation, bankruptcy legislation, capital market legislation, competition legislation, consumer protection legislation, debtor/creditor legislation, foreign trade and foreign investment legislation, insurance legislation and intellectual property legislation. Property rights are sufficiently well-designed and protected in China. Contract enforcement is sufficiently consistent to allow parties to develop and execute the valuable property exchanges. Finally, the supply of competent lawyers is growing and China's judiciary continues to improve and appears capable of enforcing the various property and contract rights that form the basis of Bayh–Dole-type technology exchanges.

5. China's intellectual property regime has come of age

> One thing necessary to stress is the need to concretely strengthen IP protection. In the new era, global science and technology competition, as well as economic competition, is primarily a competition of IP rights. Promoting IP protection therefore promotes and inspires innovation.
> – Premier Wen Jiabao[1]

IP protection provides the foundation for a university technology commercialization system. At the most general level, IP protection provides an economic incentive for individuals and firms, including universities, to invest in R&D.[2] Economic actors will not invest in the creation of technology unless they believe they will receive a profitable return on their investment. R&D can be a very expensive and time-consuming effort. To make matters more challenging, the result of R&D is not a tangible, easily protectable good. Instead, the result of successful R&D is new knowledge. By its very nature, knowledge is a partially-excludable good (i.e., it is difficult to exclude unintended parties from benefiting from knowledge),[3] which makes it difficult for inventors to collect the full value of R&D efforts. Economic historian Douglass North explains:

> Throughout man's past he has continually developed new techniques, but the pace has been slow and intermittent. The primary reason has been that the incentives for developing new techniques have occurred only sporadically. Typically, innovations could be copied at no cost by others and without any reward to the inventor or the innovator. The failure to develop systematic property rights in innovation up until fairly modern times was a major source of the slow pace of technological change.[4]

With knowledge being only a partially-excludable good, competitors are able to free-ride on the efforts of the inventor and capture some of the profits of the new knowledge – even though they did not make the costly R&D investments. This free-riding discourages R&D, because it reduces the potential returns to the inventor. IP rights, in particular patent rights, help to create a stronger economic incentive for firms and individuals to make the costly R&D investments needed to generate technological advances by making knowledge an 'excludable' good – or, at least, a more excludable good. The increased 'excludability' directs the profits from the innovation to the holder of the IP rights, and away from potential free-riders.

At a more specific level, IP protection provides the exchangeable property rights that are a prerequisite to a Bayh–Dole-type approach to technology transfer. The basic function of a Bayh–Dole-type approach to technology transfer is to place exchangeable property rights in the hands of the actors (e.g., universities) that are in the best position to develop that property and place it into the flow of commerce. This chapter examines China's IP protection system as it relates to commercializing university-developed technology and explains how China has implemented a modern IP regime that is more than sufficient to support a Bayh–Dole approach to technology transfer.

1.   CHINESE IP LEGISLATION

At the outset of the Reform Era, China did not have a functioning IP system. IP was not well protected by Chinese law and there was little acceptance by the general public of the need to protect IP.[5] In the early 1980s, however, China began a massive effort to develop IP legislation and protection that has since brought China's IP legislation in line with international standards and systems. Table 5.1 shows the steady development of China's IP legislation and multilateral agreements.

In less than 30 years, China has built a comprehensive IP protection architecture that is well aligned with international standards. While China's overall progress at developing a market-supporting legal system has been outstanding, it is possible that its progress in implementing a modern system of IP legislation has been even more impressive. One Chinese commentator recently described China's progress as follows:

> In all of Chinese legislative history, no laws have received more attention than those concerning IP. The Chinese government has tried to establish a legal system that meets the current level of IP protection in the world system. Of all the laws in China, these IP laws are the closest to corresponding laws in developed countries. In other words, China has tried, in 20 years' time, to reach the level of IP protection that it took developed countries more than 100 years to reach.[6]

1.1   China's Patent Law

Patenting is the most common form of IP protection employed by universities seeking to commercially exploit their inventions. China's first 'modern' patent law was enacted in 1984 and has been amended three times (1992, 2000 and 2008[7]) to increase the scope of protection and better

Table 5.1   Timeline of Chinese IP legislation and multilateral agreements in the reform era

Year	Action
1980	China joined WIPO
1982	Trademark Law enacted
1984	Patent Law enacted
1985	Joined Paris Convention
1989	Joined Madrid Agreement
1990	Copyright Law enacted
1991	Regulations on the Protection of Computer Software enacted
1992	● First Amendment to Patent Law to expand protection ● Joined Berne Convention
1993	● Major revisions to Trademark Law to expand protection ● First IP court established in Beijing
1994	Joined Patent Cooperation Treaty
1995	Joined Madrid Protocol
2001	● Second Amendment to Patent Law ● Second Amendment to Trademark Law ● Major revisions to Copyright Law to expand protection ● Joined WTO and TRIPS (Agreement on Trade Related Aspects of Intellectual Property Rights)
2007	● Joined WIPO Copyright Treaty ● Joined WIPO Performances and Phonograms Treaty
2008	● Third Amendment to Patent Law

align China's patent law system with international standards. The Patent Law is accompanied by a detailed set of implementing regulations (the Implementing Regulations).

1.1.1   Brief summary of China's patent law system

China's combined Patent Law statute (as amended) and regulations are comparable to developed country patent law in the types of inventions that are patentable, the level of protection provided, and in its requirements and procedures.[8]

a.   Types of patents and scope of protection   China's patent system provides for three types of patents: invention patents; utility model patents; and design patents.[9]

- *Invention Patents:*   invention patents are granted for 'any new technical solution relating to a product, a process or improvement thereof.'[10] To obtain an invention patent, the invention must be novel, inventive and have practical applicability.[11] Invention patents receive 20 years of protection from the date of filing.[12]

- *Utility Model Patents:* utility model patents are granted for 'any new technical solution relating to the shape, the structure, or their combination, of a product, which is fit for practical use.'[13] As with invention patents, the invention must be novel, inventive and have practical applicability to obtain a utility model patent.[14] Utility model patents receive 10 years of protection from the date of filing.[15]

- *Design Patents:* design patents are granted for 'any new design of the shape, the pattern or their combination, or the combination of the color with shape or pattern, of a product, which creates an aesthetic feeling and is fit for industrial application.'[16] For a design patent the invention must be 'unique', but it does not have to be novel or possess practical applicability.[17] Design patents receive 10 years of protection from the date of filing.[18]

The level of protection that a patent provides depends upon what type of patent is sought. Following the grant of an invention or utility model patent, 'no entity or individual may, without the authorization of the patentee, exploit the patent, that is, make, use, offer to sell, sell or import the patented product, or use the patented process, and use, offer to sell, sell or import the product directly obtained by the patented process, for production or business purposes.'[19] Following the grant of a design patent, 'no entity or individual may, without the authorization of the patentee, exploit the patent, that is, make, sell or import the product incorporating its or his patented design, for production or business purposes.'[20]

b.   Patentable subject matter   In addition to the novelty, inventiveness, practical applicability and uniqueness requirements set forth in Articles 22 and 23, China's Patent Law also expressly excludes certain categories of invention from patent protection. For example, patent rights may not be granted for scientific discoveries,[21] although technical solutions that come from a scientific discovery may still be patentable.[22] China's Guidelines for Patent Examination (the Guidelines) provide the following explanatory example:

[D]iscovery of the photosensitive property of a silver halide under illumination cannot be granted a patent right. However, a patent right may be granted for the photographic film and the process to produce the film in accordance with this discovery.[23]

China's Patent Law also excludes rules and methods for mental activities from patent protection.[24] The Guidelines explain that rules and methods for mental activities are not patentable because they do not involve any technical characteristics.[25] The Guidelines go on to provide a long list of activities that are considered to be rules and methods for mental activities, including mathematical theories, methods of teaching, training and presenting, and 'computer programs *per se*.'[26] See Box 5.1 for a discussion on the availability of 'software patents' in China. Methods of doing business are presumably treated as rules and methods for mental activities and are, therefore, not patentable.[27]

Methods for the diagnosis or for the treatment of diseases, animal and plant varieties and substances obtained by means of nuclear transformation are also excluded from patent protection under China's Patent Law.[28] Finally, China's Patent Law refuses patent protection for inventions that are contrary to China's laws or social morality or that violate public policy.[29]

c. Filing and application process   Like most European countries, China operates under a 'first-to-file' rule,[30] which provides the patent grant to the first applicant to file a patent application with the patent office irrespective of whether the first one to file was the initial creator of the invention. China's Patent Law also provides for both international and domestic priority.[31] China's application and examination process has greatly evolved over the last 20 years and those processes appear to comply with international standards. Because of the greater rights afforded to their holders, the examination process for invention patents is longer and more detailed than for utility model and design patents. Invention patents must undergo a substantive examination,[32] while utility model and design patents only undergo a preliminary examination.[33]

1.1.2   Key Patent Law provision for China's Bayh–Dole system: 'inventions for hire'

Chinese Patent Law includes an 'invention for hire' doctrine. As a general rule, the creator of an invention has the right to apply for a, and own the resulting, patent for that invention. Under Chinese law, where the inventor is an employee, the employer will have the right to apply for a patent – and, if approved, will own the patent – if the invention was made by the employee:[34]

1. Within the scope of her employment, or otherwise in execution of a task for the employer;
2. Within one year from her resignation, retirement or change of work,

BOX 5.1 AVAILABILITY OF 'SOFTWARE' PATENTS

While 'computer programs *per se*' are expressly included on the list of non-patentable rules and methods for mental activities, the ability to obtain 'software' patents is a bit more subtle due to the definition of 'computer programs *per se*' in the Guidelines and how it is differentiated from 'computer program-related inventions.'

> *Computer programs per se* = 'a coded instruction sequence which can be executed by a device capable of information processing, e.g., a computer, so that certain results can be obtained, or a symbolized instruction sequence, or a symbolized statement sequence, which can be transformed automatically into a coded instruction sequence. Computer programs per se include source programs and object programs.'[1]

Computer program-related inventions can be differentiated from computer programs per se. Wenping Chen & Xun Feng wrote an article on patentable subject matter under Chinese Patent Law that provides the following explanation:

> The computer program-related inventions refer to the solutions that are completely or partially based on the execution of computer program to solve the problem raised in the inventions. Therefore, a computer program *per se* can be understood as an 'expression' in the form of an instruction sequence of the 'idea' (solution) underlying the program. The exclusion of computer programs *per se* is limited to the software product as such. The patentability of the underlying 'solution' depends on whether it is a technical solution or not. When the solution underlying a computer program solves a technical problem, utilizes technical means or produces technical effects, it possesses technical characteristic(s) and constitutes a technical solution that may qualify for a patent under [China Patent Law], even if it is entirely based on a computer program.[2]

Notes

1. Chapter 9, Section 1 of the Guidelines,
2. Wenping Chen & Xun Feng (2003), 'The China IP Focus 2003: How to Distinguish Patentable Subject Matter,' *Managing Intellectual Property* (January), available at http://managingip.com/Article.aspx?ArticleID=473279.

BOX 5.2   GUIDELINES ON REASONABLE REMUNERATION FOR INVENTIONS FOR HIRE

The guidelines provide that, unless otherwise agreed between the employee and the employer, inventors are entitled to compensation at three stages:

1. Issuance of patent: within three months of the patent's issuance, the employer shall pay the employee/inventor a minimum of 3000 RMB for an invention patent and a minimum of 1000 RMB for a utility model or design patent.[1]
2. Patent exploited by employer: for the duration of the patent, the employer shall pay the employee/inventor at least 2% of the after-tax profits generated by an invention or utility model patent, or 0.2% of the after-tax profits generated by a design patent. Such payment can be made on an annual basis or a lump sum payment.[2]
3. Employer licenses patent to a 3rd party: the employer shall pay the employee/inventor at least 10% of the after-tax profits generated by licensing the patent.[3]

Notes

1. Rule 77 of the Implementing Regulations.
2. Rule 78 of the Implementing Regulations.
3. *Id.*

---

where the invention relates to the employee's former scope of employment or other tasks assigned by the employer; or
3. Mainly by using the employer's materials (including the employer's money, equipment, spare parts, raw materials or technical materials which are not disclosed to the public).

Under U.S. law, an employer can obligate his/her employees to assign inventions for hire without having to pay any additional compensation to the employee. In China, the Patent Law requires that an employer pay 'reasonable remuneration' to an employee/inventor for the transfer, exploitation and licensing of an invention for hire.[35] The Implementing Regulations for the Patent Law, which went into effect on February 1, 2010,[36] provide guidelines for this reasonable remuneration (see Box 5.2).

The guidelines on reasonable remuneration are just that . . . they are 'guidelines.' Employers are entirely free to compensate their employees more generously for their inventions for hire. Based on interviews we conducted with the technology transfer directors for a few of China's elite research universities, university scientists at some universities are compensated at a much higher level than what is called for in the guidelines – e.g., they receive 50% to 70% of the profit from commercialized inventions. Such a result is not at all surprising as Chinese universities should be expected to experiment with different reward strategies to find the most efficient incentive structure for their particular institution.

For employee inventions that do not qualify as inventions for hire, the right to patent the invention rests with the employee.[37] A key issue for universities interested in technology transfer is whether an invention qualifies as an invention for hire. A 1999 presentation by Robert Kneller on university inventions in Japan and China explained that, in 1999, universities were not aggressively asserting their rights over inventions for hire:

> One of the principal issues in China is determining whether an invention is [an invention for hire]. Before universities established [technology transfer offices] and before they had an effective technology transfer system, there were great incentives to classify inventions as non-[inventions for hire]. This way they could easily be passed to companies and the inventors could realize some reward from the companies . . .
>
> Even today an inordinate percentage of inventions are classified as non-[inventions for hire]. In Tsinghua University, roughly China's MIT, about 40% of inventions are classified as non-[inventions for hire]. In Xian Communications University, this percentage is said to be considerably higher. This reflects the carry over of practices from the former system, the independent mindedness of faculty members, and a skepticism by faculty concerning the effectiveness of technology transfer via the new [technology transfer offices]. However, universities are asserting their authority to make the final decision concerning whether inventions are [inventions for hire].[38]

Kneller's observations are seriously dated, but they do provide a reminder that universities must be vigilant about managing and protecting their IP rights. Chinese Patent Law provides that if an employer and employee enter into a contract specifying the patent application/ownership rights for employee inventions that are made with the employer's materials or technical means, that contract will govern.[39] For Chinese universities seeking to employ a Bayh–Dole strategy, it is highly advisable that they enter into contractual agreements with their university researchers that cover the patent application/ownership rights of any resulting inventions and thereby avoid any debate over whether an invention is an invention for hire.

1.2 China's Copyright Law

In addition to seeking patents, universities may also choose to rely on copyrights as a protection strategy for commercializing certain types of university inventions, such as software programs. China's Copyright Law was enacted in 1990, and substantially amended in 2001 to bring the Copyright Law into accord with both TRIPS and the Berne Convention. China's Copyright Law, which is administered at the national level by the national Copyright Administration Department (operating under the State Council) and at the local level by provincial offices of the Copyright Administration Department,[40] consists of China's Copyright Law and a set of implementing regulations, including regulations implementing international copyright treaties and rules and regulations protecting software programs.

1.2.1 Works for hire

For university technology transfer, 'work for hire' is one of the more relevant doctrines. Under traditional principles of copyright law, the creator of a copyrightable work will own the copyright. In many countries, including the United States, the work for hire doctrine changes that general rule by granting the copyright to the party who pays for the work's creation, rather than to the creator of the work. In the employment setting that means the employer will be the copyright owner of a work for hire unless otherwise agreed by the parties. China takes a different approach to the work for hire doctrine. Under Chinese Copyright Law, the copyright to a work for hire belongs to the employee/author, not the employer, subject to a number of important limitations and exceptions.[41]

a. Limitations on the employee's copyright While employees hold the copyright in a work for hire, employers retain a number of important rights in such works. First, the employer is given a priority right to use the employee's work for hire within the employer's normal scope of business.[42] Second, within two years of the completion of the work, the employee cannot authorize a third party to exploit the work in the same way as the employer unless the employee has obtained the employer's consent.[43]

b. Exceptions that provide copyright to employer Under either of the following circumstances, the employer will hold the copyright to a work for hire, although the employee will retain the moral rights of authorship:

1. Engineering design drawings, product design drawings, maps, computer software and other works created in the course of employment mainly with the material and technical resource of the legal entity or other organization and under its responsibility.[44]
2. Works created in the course of employment where the copyright is assigned to the employer by law, administrative regulation, or a contract between the employer and employee.[45]

As with patents, Chinese universities should consider entering into agreements with their professors and researchers that specify the university's copyright ownership for works for hire.

1.2.2 Regulations on Computer Software Protection – software for hire

Software can frequently be relevant to university technology transfer. In addition to the potential patent protection for computer-related inventions that was discussed above (see Box 5.1), software also enjoys copyright protection under China's Copyright Law and the Regulations on Computer Software Protection (the Software Regulations) that were promulgated by the State Council in December 2001 pursuant to Article 58 of China's Copyright Law.[46] The Software Regulations employ a 'software for hire' doctrine that is more comparable to China's 'invention for hire' doctrine than to its 'work for hire' doctrine. Where an employee develops software for his employer, the employer will own the copyright in the software if:

*   The software is developed based on objectives explicitly designated by the employer;
*   The software is a foreseeable or natural result of the employee's activities in the scope of employment; or
*   The software is developed mainly with the material and technical resources of the employer (e.g., funds, special equipment or unpublished special information) and the employer assumes the responsibility for the software.[47]

2.   IP ENFORCEMENT IN CHINA

It has become standard fare for commentators to observe that China has succeeded in adopting laws and regulations that provide IP owners with rights and protections that are reasonably aligned with international norms, but that China continues to struggle with the actual enforcement of those rights and protections.[48] Statements that China's IP enforcement is weak are almost never challenged. Walk into any meeting of Western business people, politicians, IP lawyers, or academics and state 'Chinese IP enforcement is weak' – and invariably the pronouncement will be received with resounding approval from the audience. For many, being told that Chinese IP enforcement is weak is like being told that the sky is blue – it is a truism that requires little thought or analysis. There is a potential problem, however, with such broad pronouncements of Chinese IP enforcement failures – they may be wrong.

2.1 Is China's IP Enforcement Really that Weak?

In order for IP rights to be valuable, holders of the rights must be able to enforce them. An effective IP enforcement system is needed to ensure that the rights provided by IP legislation are meaningful. Just as we saw with the evolution of China's legal system, however, Western perceptions of China's IP enforcement strength frequently fail to appreciate the meaningful progress that is taking place.

2.1.1 Counterfeit and pirated goods

When critics charge China with weak IP enforcement, the complaints typically focus on China's problems with counterfeit and pirated goods. On the counterfeiting side, critics point to the incredible number of fake goods that are produced in China each year. From knock-off luxury goods (e.g. fake Rolexes, fake Coach bags, and fake designer clothes and shoes) to more advanced counterfeit products (e.g., pharmaceuticals, car parts, and computer chips), the production and distribution of counterfeit goods and products has been a major problem in China for quite some time. The counterfeiters are infringing on the trademark of another and are attempting to steal some of the goodwill that the rightful trademark owner has created in its mark. On the piracy side, the IP infringement complaints typically focus on copyright infringement that takes place in China. Illegal copies of software, music, movies, and books are all significant problems in China.

Counterfeiting and piracy problems within a country do not lend themselves to reliable data collection, which means there are no official statistics on the magnitude of the problem. This data void is frequently filled by various industry associations. Unfortunately, these industry associations tend to benefit from generating a great deal of concern over the, counterfeiting/piracy problem, so their data should be taken with a grain of salt.[49] Table 5.2, for example, presents software piracy data that was recently published by the Business Software Alliance.[50]

We are in no way suggesting that China does not have a serious counterfeiting/piracy problem. We believe it does have a serious problem.

Table 5.2   Software piracy in select countries (2004–2008)

	2004	2005	2006	2007	2008
Select Lower Piracy Countries					
United States	21%	21%	21%	20%	20%
Japan	28%	28%	25%	23%	21%
United Kingdom	27%	27%	27%	26%	27%
Germany	29%	27%	28%	27%	27%
South Africa	37%	36%	35%	34%	35%
Singapore	42%	40%	39%	37%	36%
Taiwan	43%	43%	41%	40%	39%
France	45%	47%	45%	42%	41%
South Korea	46%	46%	45%	43%	43%
Select Higher Piracy Countries					
Georgia	n/a	n/a	n/a	n/a	95%
Bangladesh	n/a	n/a	92%	92%	92%
Vietnam	92%	90%	88%	85%	85%
Nigeria	84%	82%	82%	82%	83%
China	90%	86%	82%	82%	80%
India	74%	72%	71%	69%	68%
Russia	87%	83%	80%	73%	68%

Source: Business Software Alliance (2009), *08 Piracy Study – Sixth Annual BSA-IDC Global Software*, pp.12–13, available at http://global.bsa.org/globalpiracy2008/studies/globalpiracy2008.pdf.

The level of the problem, however, may be overstated.[51] Moreover, those critiquing China's counterfeiting/piracy problem tend to omit the myriad of efforts that China's government has pursued to try and lower the counterfeiting/piracy problem to a more acceptable rate. Are they trying to imply that China's government does not care that counterfeiting/piracy is taking place, or, worse yet, wants it to take place? While we are unclear on the motivations for critics' frequent failure to point out the Chinese government's effort to combat counterfeiting/piracy, we do believe the government's effort is a sincere attempt to reduce the problem. Table 5.3 provides select statistics for some of the efforts undertaken by China's trademark enforcement agency (the Administration for Industry and Commerce (AIC)), copyright enforcement agency (National Copyright Administration of China (NCAC)), and customs officials. In addition to the efforts of its administrative agencies, China's courts have become much more active in resolving criminal IP cases (Table 5.4).

Table 5.3 *Select anti-counterfeiting/piracy efforts by China's AIC, NCAC and customs officials*

	2004	2006	2008
Trademark Enforcement by the AIC			
Investigations of trademark violations	51851	50534	56634
No. that were for trademark infringement/ counterfeiting	40171	n/a	47045
No. that involved foreign-related trademarks	5494	9562	10965
Seizures and removals of sets of illegal trademark labels	39.0 mm	30.4 mm	19.6 mm
Fines imposed	RMB 268 mm	RMB 398 mm	RMB 467 mm
Cases turned over to law enforcement	96	252	137
Copyright Enforcement by the NCAC			
Efforts to promote use of licensed software in enterprises			
Large enterprises listed as users of licensed software	n/a	n/a	1500
Other enterprises listed as users of licensed software	n/a	n/a	7600
Crackdown on Internet piracy (e.g., illegal downloading of movies, music, software and books)			
Websites ordered to delete or block infringing content	n/a	361	173
Websites shut down	n/a	205	192
Cases turned over to law enforcement	n/a	6	10
Customs Seizures			
Seizures of imported or exported infringing goods	n/a	200 mm pieces	600 mm pieces

Sources: Source for 2008 statistics: SIPO, *White Paper on China's Intellectual Property Protection in 2008* [hereinafter 2008 SIPO IP Protection White Paper]. Source for 2006 statistics: SIPO, *White Paper on China's Intellectual Property Protection in 2006* [hereinafter 2006 SIPO IP Protection White Paper]. Source for 2004 statistics: SIPO, *White Paper on China's Intellectual Property Protection in 2004* [hereinafter 2004 SIPO IP Protection White Paper]. Each of these white papers is available at www.sipo.gov.cn/sipo_English/laws/whitepapers/.

Table 5.4   Increase in China's criminal IP caseload

	2004	2006	2008
Criminal IP Cases			
Total no. of criminal IP cases resolved by Chinese courts	2753	2277	3326
No. that involved criminal violation of IP rights (e.g., trademark infringement)	387	769	996
No. that involved manufacture and distribution of goods with inferior quality (in violation of IP rights)	932	437	610
No. that involved illegal business operations (in violation of IP rights)	1434	1066	1707
Other	–	5	13

Sources: Source for 2008 statistics: *2008 SIPO IP Protection White Paper*. Source for 2006 statistics: *2006 SIPO IP Protection White Paper*. Source for 2004 statistics: *2004 SIPO IP Protection White Paper* all available at www.sipo.gov.cn/sipo_English/laws/whitepapers/.

Another significant problem that weakens the analysis of many critics of China's IP enforcement strength is the failure to appreciate China's regional disparities. China is not a homogenous country, but instead is a large, complex and diverse country with substantial disparities between its various regions.[52] China's eastern regions were the first to be opened up in the 1980s, and they have enjoyed the lion's share of China's economic progress since that time.[53] Disparities between China's regions continue to grow and have become a major focus for China's government.[54] These regional disparities caused noted Chinese economist Hu Angang to characterize China in 2003 as being 'one China, four worlds,'[55] to reflect that:

1.   Roughly 2 per cent of China's population lives in modern cities like Beijing, Shanghai and Shenzhen that have reached developed-economy income levels (the First World);

2.   Twenty-two per cent live in China's coastal provinces, e.g., Guangdong, Jiangsu, Liaoning and Zhejiang, whose income levels are comparable to those of upper-middle-income countries (the Second World);

3.   Twenty-six per cent live in Hebei, Heilongjian, Hubei, Jilin and some of China's central provinces that are comparable to lower-middle-income countries (the Third World); and

4.   Roughly half the population remains in lower-income country conditions (the Fourth World).[56]

These regional disparities can have a profound impact on China's counterfeiting/piracy problem. As a general rule, the strength of IP enforcement for developing countries improves as the country's income level rises.[57] China appears to be following that pattern, but is doing so at a regional level. As Professor Peter Yu explains:

> While stronger intellectual property protection and the emergence of intellectual property-based industries in Beijing, Shanghai, Guangzhou, and other major cities and coastal regions have led to greater improvement of protection in the affected places, piracy and counterfeiting have not migrated out of the country. Instead, they have spread to other parts of the country, whose conditions are no different from those of the big cities a decade ago when intellectual property protection began to strengthen. In light of this migration, intellectual property problems are likely to remain in the country in the near future, even if some of these problems have migrated to countries in Southeast Asia.[58]

Overall, it appears that China is making significant strides in combating counterfeiting and piracy problems in its so-called First World and Second World, but that China's Third and Fourth Worlds continue to have more severe counterfeiting and piracy problems.

2.1.2 What do China's counterfeiting/piracy problems have to do with university technology transfer?

As we explained above, critics tend to use China's counterfeiting/piracy problems as a proxy for the country's overall IP enforcement strength. The typical argument is made roughly as follows: because a disproportionate amount of the world's knock-off consumer goods and copyright infringement is taking place in China, China's ability to enforce IP rights must be weak. Is that a fair connection to make? It turns out that with regards to China's Bayh–Dole system, it is not a fair connection to make at all. China's counterfeiting and piracy problem is only tangentially relevant to China's Bayh–Dole system.

At first glance, the minimal relevance of China's counterfeiting/piracy problem to its Bayh–Dole system may seem counterintuitive. The whole purpose of a Bayh–Dole system is to strengthen the IP rights of universities – to incentivize the universities to develop and commercialize technology – and the strength of IP rights is highly dependent on the country's IP enforcement system. If that is all true, and it is, how can China's counterfeiting/piracy problem not create a drag on China's Bayh–Dole effort? The reason is that China's counterfeiting/piracy problems are primarily trademark and copyright problems, not patent problems. For example, the United States' widely-reported WTO complaint against China for insufficient IP protection and enforcement focused on trademark and copyright issues, not patent issues.[59] While trademark and copyright are serious problems for a country to address, most university-generated technology will come in the form of highly-sophisticated, patented-technology, which makes China's counterfeiting/piracy problems a bit of a red herring. Problems with knock-off consumer products and illegal copies of copyrighted materials have little relation to the IP that China's Bayh–Dole system seeks to incentivize at its universities. The most critical IP enforcement for a Bayh–Dole system tends to be patent enforcement – namely, the ability to enforce patent rights. As with any country, China is not immune from patent piracy. However, China's clichéd reputation as a pirate nation[60] stems from its considerable problems with trademark and copyright infringement rather than patent infringement.

If patent enforcement is the primary issue for China's Bayh–Dole system, and China's greatest counterfeiting/piracy problems relate to trademarks and copyrights, why have we spent so much time in the section addressing it? We are addressing this issue because we are concerned that negative perceptions about the strength of China's IP enforcement, even if incorrect, can significantly impact the behavior of innovation system actors. While the adage from the Andre Agassi Canon commercials from the early 1990s – 'image is everything' – may not be entirely accurate, perceptions do matter a lot for IP enforcement. IP enforcement is not simply a matter of government enforcement. As in most countries, effective IP enforcement in China requires the IP owners to proactively pursue IP infringers,[61] not simply sit back and hope that law enforcement officials will take care of things. When an IP owner believes its IP rights are being infringed in China, that owner must aggressively seek to protect its rights through China's various IP enforcement mechanisms. If, however, IP owners believe that IP enforcement is weak – and we are concerned that many patent owners fail to appreciate that China's high profile trademark and copyright infringement problems are a red herring for their patent enforcement – they may fail to aggressively protect their IP rights because they believe such efforts would be a waste of time and resources.[62] Failing to aggressively protect IP rights can create a vicious cycle that promotes continuously weaker IP enforcement (see Figure 5.1).

We believe that it is important, therefore, for Chinese policymakers and others interested in strong Chinese IP rights to break this vicious cycle and work to clearly communicate to IP owners the improved mechanisms for enforcing IP rights and the need for such owners to aggressively protect those IP rights (see Figure 5.2).

This book, and this chapter, is focused primarily on Chinese patent rights and patent protection as they relate to university technology transfer. However, the need to break the vicious cycle caused by overly negative

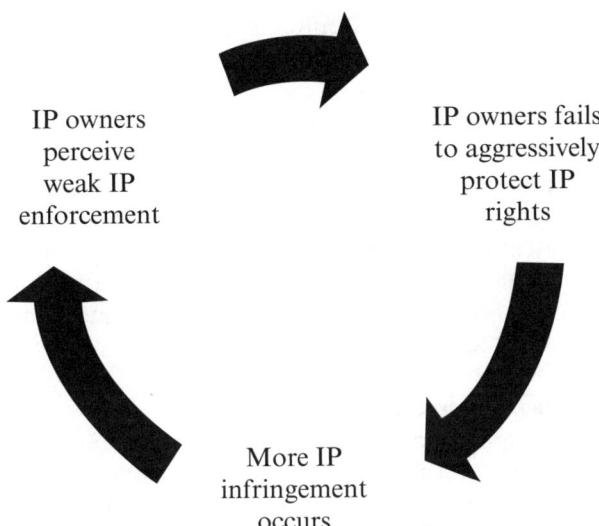

Figure 5.1   Perception of weak IP enforcement creates a vicious cycle

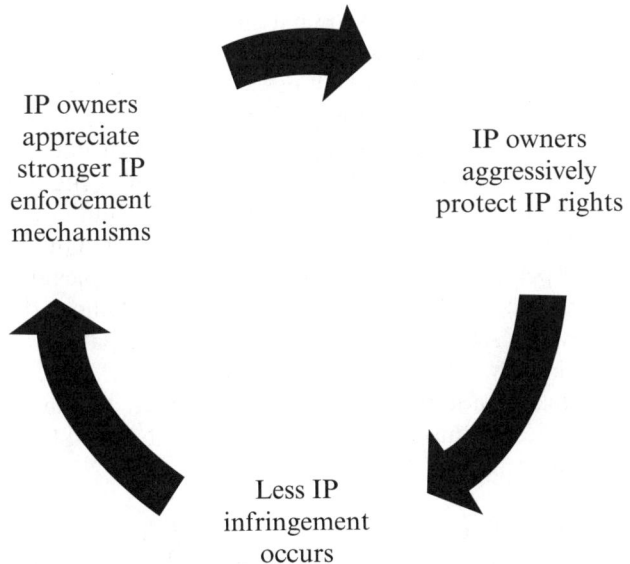

Figure 5.2   Better understanding of IP enforcement system creates a virtuous cycle

perceptions of China's overall IP enforcement is a crucial issue for China's policymakers to address.

2.2   Can Parties Effectively Enforce Patents in China?

Since patents are the primary IP rights that result from a Bayh–Dole system, this section will examine whether patent owners can effectively enforce their patent rights in China. China provides a mix of governmental enforcement mechanisms and private remedies that a patent owner can employ to enforce its patent rights. Collectively, these mechanisms provide a reasonable set of tools for IP enforcement.

2.2.1   Government enforcement actions

The Chinese government has broad authority to remedy and sanction patent infringement. The government can take administrative action through various administrative agencies or it can pursue criminal actions. In each case, patent owners should not simply rely on the administrative agencies or law enforcement officials to discover and prosecute infringement cases on their own. While government actors have the authority to investigate and bring actions on their own, the practical reality is that they will seldom act until a complaint is filed by the patent owner coupled with evidence that a law or regulation has been violated.[63] If the government decides to take action to protect patent rights, it possesses a full array of enforcement tools.[64] Table 5.5 provides a summary of those enforcement tools.

Administrative actions tend to be the primary government IP enforcement tool. Criminal law enforcement actions are generally reserved for the 'large-scale or repeat cases of commercial counterfeiting.'[65] The primary administrative agency for patent infringement claims is the State Intellectual Property Office (SIPO). Complaints are typically brought at the local level – i.e., at local Intellectual Property Offices (local IPOs). Table 5.6 provides select statistics for some of the patent enforcement efforts undertaken by SIPO and local IPOs since 2004.

One drawback of administrative IP enforcement actions that is frequently cited is local collusion or corruption. John Lee and Eion Murdock of China's Lehman, Lee & Xu law firm, explain:

> [T]he jurisdiction of administrative agencies is typically quite local in scope, and multiple complaints may be required when dealing with infringement on a regional or national level. A further problem with administrative enforcement is the risk that the [local] administrative agency may collude with the infringer or simply refuse to act. Collusion is likely to be more of a problem in smaller cities where the infringer has developed a strong relationship with the local authorities.[66]

Table 5.5   Basic government IP enforcement tools

Tool	Description
Investigations	Officials have the ability to conduct on-site inspections and raids, question potential infringers under oath and subpoena witnesses.
Injunctions	Officials can seek temporary or permanent injunctions against infringers. Injunctions are issued by Chinese courts.
Confiscate infringing goods	Officials can temporarily seize allegedly infringing goods until a proceeding determines whether such goods do infringe another party's IP rights.
Disgorge profits and pay monetary fines	Officials can require IP infringers to disgorge any ill-gotten profits and pay additional monetary fines. Money collected by government officials, however, remains with the government and is not paid to the wronged IP owner.
Criminal penalties	Law enforcement officials may seek prison time for IP offenders. Prison sentences may only be issued by China's criminal courts.

Sources: Chapter VII of China's Patent Law and John Lee & Eion Murdock (2009), 'IP Enforcement in China', *Supplement – China IP Focus 2009* (7th ed.), available at www.managingip.com/Article/2176060/IP-enforcement-in-China.html.

Table 5.6   Select statistics of patent enforcement efforts by SIPO and local IPOs

	2004	2006	2008
Patent infringement disputes accepted	1455	1227	1092
Patent infringement disputes concluded	1215	973	660
Patent counterfeit cases	345	973	59
Commercial premises inspected and investigated for infringing patents	n/a	7780	7671
Cases turned over to law enforcement	n/a	12	21
Joint law enforcement actions	n/a	469	327

Sources: Source for 2008 statistics: 2008 SIPO IP Protection White Papers; source for 2006 statistics: 2006 SIPO IP Protection White Paper; source for 2004 statistics: 2004 SIPO IP Protection White Paper, all available at www.sipo.gov.cn/sipo_English/laws/whitepapers.

Table 5.7   IP cases handled by China's civil courts

	2004	2006	2008	% increase 2004–2008
Civil IP Cases – overall statistics				
Trial cases heard	9329	14219	24406	162%
Trial cases resolved	8332	14056	23518	182%
Appellate cases heard	n/a	2686	4759	n/a
Appellate cases resolved	n/a	2652	4699	n/a
Retrials ordered	n/a	42	102	n/a
Retrials resolved	n/a	42	71	n/a
Civil Trial Cases – by type of IP infringement				
Patent cases heard	2549	3196	4074	60%
Trademark cases heard	1325	2521	6233	370%
Copyright cases heard	4264	5719	10951	157%
Licensing cases heard	n/a	668	623	n/a
Unfair competition cases heard	n/a	1188	1185	n/a
Other IP infringement cases heard	n/a	844	1340	n/a

Sources: Source for 2008 statistics: 2008 SIPO IP Protection White Paper; source for 2006 statistics: 2006 SIPO IP Protection White Paper; source for 2004 statistics: 2004 SIPO IP Protection White Paper, all available at www.sipo.gov.cn/sipo_English/laws/whitepapers.

To be more precise, we would expect collusion/corruption to be most problematic in China's least developed regions – i.e., China's Third and Fourth Worlds – where IP rights have yet to become central to their economic activities and growth.

2.2.2   Private remedies

In addition to government actions, China's IP regulatory system also recognizes the importance of private actions for addressing IP infringement and provides a comprehensive set of private remedies. These private actions form an integral part of China's overall IP enforcement scheme. In developed countries, private legal actions are the dominant enforcement mechanism for combating patent infringement. With the significant improvement of China's judicial system (see Chapter 4), private actions have begun to play a critical patent enforcement role in China. Table 5.7 provides summary statistics on the ever increasing role of China's civil judiciary in resolving IP disputes.

On a purely practical note, IP owners need to understand that the IP competence of China's courts is far from uniform when deciding whether (or where) to bring a private lawsuit.[67] That means that forum selection is particularly important for patent litigation in China. Fortunately, China's venue rules provide plaintiffs with a fair amount of flexibility in where they may bring their lawsuits.

2.3 Lessons from the Viagra Patent Dispute in China

One of the most eventful, and illustrative, examples of patent enforcement in China is Pfizer's patent covering sildenafil citrate, which is the key ingredient involves Pfizer's internationally famous erectile dysfunction drug, Viagra (the Viagra patent).[68] The history of the Viagra patent in China spans a period of almost 15 years that has witnessed considerable growth in the sophistication and breadth of patent protection in China.

2.3.1 Background of the Viagra patent
Pfizer filed its initial patent for the Viagra compound in the United Kingdom in June 1990.[69] Based on this U.K. filing, Pfizer also obtained patents in the United States, the European Union, Japan and a number of other countries.[70] In each case, the patents sought to protect the use of sildenafil citrate as a treatment for various cardiac and circulatory disorders, but made no mention of its use for treating erectile dysfunction.[71] Interestingly, Pfizer could not file a patent application in China based on its 1990 U.K. patent application because China's patent law did not protect pharmaceutical compositions until 1993 when China joined the Patent Cooperation Treaty.[72] Pfizer soon discovered that the optimal use for Viagra was treating male erectile dysfunction, and filed a U.K. patent application to protect that use in June 1993. This application was filed in China through the PCT process and Pfizer eventually obtained the Viagra patent in China on September 19, 2001 with a single claim:[73]

> The use of 5-[2-ethoxy-5- (4-methyl-1-piperazinylsulphonyl)-phenyl]-1-methyl-3-n-propyl-1,6-dihydro-7H-pyrazolo94,3-d] pyrimidin-7-one or of a pharmaceutical composition containing any of the same, for manufacture of a medicament for curative or prophylactic treatment of erectile dysfunction in a male animal, including man.[74]

Pfizer began to experience problems with its Viagra patent in 1998. Revocation petitions were filed against the Viagra patent in the European Patent Office in 1998 and in the United Kingdom in 1999, and the patents were revoked in those jurisdictions in 2000 'for lack of inventive step' (i.e., the claims failed to show novelty).[75] Pfizer has since faced numerous challenges to its Viagra patent in countries around the world, and has had the patent invalidated in a number of countries including Australia[76] and several South American countries.[77]

2.3.2 Chinese Viagra patent invalidated

Against this backdrop, twelve Chinese companies, and one individual, filed invalidation petitions against the Viagra patent on the first day it was granted in China, September 19, 2001.[78] Article 45 of China's Patent Law allows for private parties to file invalidation petitions, and does not impose any particular standing requirement on such private parties.[79] SIPO's Patent Reexamination Board (PRB) has exclusive jurisdiction to decide patent invalidation petitions.[80] The PRB announced its decision on July 5, 2004, and decided to invalidate the Viagra patent.[81] A written decision was never issued, but various reports exist of PRB statements made following the decision.[82] Those statements indicate that the Viagra patent was invalidated because Pfizer did not satisfy Article 26 of China's Patent Law, which provides in pertinent part:

> Where an application for a patent for invention or utility model is filed, a request, a description and its abstract, and claims shall be submitted. . . . *The description shall clearly and completely describe the invention or utility model so as to enable a person skilled in the relevant field of technology to carry it out . . .* [emphasis added][83]

The PRB found that the technical descriptions in the Viagra patent were insufficient to confirm that sildenafil citrate could relieve erectile dysfunction. In short, 'the invention could not be reproduced by experts based solely on the information provided in the patent application.'[84] Interestingly, the PRB chose not to invalidate the patent on the novelty argument that was relied on by the European and U.K. authorities.

2.3.3 International outcry against China

The international outcry following SIPO's invalidation of the Viagra patent was highly critical of China's commitment to IP protection and many charged China with giving preferential treatment to local companies. Some even suggested that the Viagra invalidation could violate China's WTO commitments – specifically Article 27(1) of the TRIPS Agreement which requires member countries to protect patent rights 'without discrimination as to the place of invention, the field of technology and whether products are imported or locally produced.'[85] Naotaka Matsukata, the former director of policy planning for the U.S. Trade Representative, provided one of the more heated criticisms of the decision:

Faced with rising global pressure to crack down on patent infringement, Beijing may be in the process of redefining patent criteria effectively to safeguard Chinese drug-makers from accusations of illegal infringements. The removal of patents on Viagra . . . would offer Chinese companies free rein to manufacture homegrown copycat drugs without fear of prosecution. . .

For the global research pharmaceutical industry, the ruling carries the significant threat of a Chinese government tacitly supporting the production of counterfeit drugs by domestic Chinese companies. For China's trading partners worldwide, the ruling demonstrates China's somewhat cautious embrace of the WTO's rules-based system, which it joined in 2002.[86]

Despite their vehemence and frequency, the validity of these criticisms against SIPO and China's overall IP system was highly questionable. China may have been the only country to invalidate the Viagra patent on disclosure grounds, but it was by no means the only country to challenge the patent or even to invalidate it. And yet, many commentators singled out China as a rogue, pirate nation for its actions. China's historical problems with IP enforcement were a likely reason that it took so much criticism. Few external commentators were willing to stand up and explain that China possessed a legitimate patent protection system, so those who decried SIPO's Viagra decision with skepticism and derision largely went unchallenged. China was not completely without fault, as the lack of transparency around the Viagra decision – i.e., the PRB's failure to publish a written opinion – did not help China's case. Failing to promptly publish a written opinion reinforces critics' charges that the decision was not based on sound legal reasoning, but instead was a political decision to support Chinese pharmaceutical companies.

It is worth noting that some commentators did not view the Viagra invalidation as an indictment of China's commitment to IP protection, but instead considered it a positive sign of the growing importance of patents and the patent protection system in China's economy.[87] Professor Peter Yu has been one of the more vocal of these commentators. Professor Yu explained:

[I]t was the first time Chinese companies took the legal route to challenge a patent owned by a major foreign company. A decade ago, local companies simply ignored the law and manufactured counterfeit products; many still do today. This time, however, local companies went to the patent office first, asking for the cancellation of Pfizer's patent . . . That is a great improvement and is largely due to the legal reforms introduced in the wake of the WTO accession.[88]

We tend to agree with Professor Yu's characterization of the matter. What was most important about the Viagra dispute was not the outcome, which itself was not entirely unreasonable, but the fact that Chinese companies were embracing the legal system as an integral part of their commercial strategy.

2.3.4 Chinese courts reinstate the Viagra patent

Despite the media attention generated by Pfizer's high-profile loss in front of the PRB, Pfizer never actually lost its patent protection for Viagra in China. The PRB does not have final say on the validity of a patent in China. Article 46 of China's Patent Law allows a patentee – in this case Pfizer – the right to appeal PRB decisions to China's court system.[89] Pfizer filed an appeal on September 28, 2004 with the Beijing No. 1 Intermediate People's Court to challenge the PRB's invalidation of the Viagra patent.[90] The appeal blocked Chinese generic-drug companies from distributing Viagra because it prevented them from getting the necessary marketing approval for their generic versions of Viagra.[91] Chinese patent lawyer Tony Chen explains:

> [A] Chinese patent is treated as valid until the invalidation decision has become final and nonappealable, and the State Food and Drug Administration ('SFDA') of China will not grant marketing approval to generic drugs while a valid patent exists for the original product.[92]

So far, the Viagra story in China has a happy ending for Pfizer. On June 2, 2006, Pfizer won its appeal with the Beijing No. 1 Intermediate People's Court, at which point the Viagra patent was reinstated.[93] Ten of the 13 original petitioners appealed this decision to the Beijing High People's Court, which also ruled in favor of Pfizer on September 7, 2007.[94] The decision by the High People's Court puts an end to any claims that the Viagra patent is invalid due to insufficient disclosure.[95] We say the story has a happy ending 'so far,' because the petitioners may continue to pursue invalidation of the Viagra patent due to lack of novelty or some other deficiency. But, as of the publishing of this book, the Viagra patent remains valid in China.

2.3.5 Does the Viagra dispute provide any useful lessons about patent protection in China?

Overall, the Viagra dispute provides a very complimentary example of the enormous advances in patent protection that have taken place in China. Each of the major actors in the saga acted in a way that shows a strong commitment to patent protection, specifically, and the rule of law, generally.

- *The Petitioners:* as we explained earlier, the Chinese pharmaceutical companies that challenged Pfizer's Viagra patent proactively embraced the legal system as they developed their commercial strategy to sell generic Viagra in China.

- *The PRB:* despite its incorrect conclusion, the PRB performed its function of reviewing the Viagra patent.
- *The Court System:* China's court system was able to correct the PRB's invalidity decision. In doing so, the courts were able to develop meaningful judicial precedent regarding Article 26 of China's Patent Law.
- *Pfizer:* Pfizer was able to demonstrate the need for patent holders to aggressively defend their patents and the benefits that can accrue from such efforts.

As more petitioners use the legal system to challenge patents and more patent holders aggressively defend their rights, enforcement of patent rights in China should 'visibly' improve[96] and could, in the not too distant future, eventually overshadow China's pirate reputation. Such efforts help to increase the sophistication of China's patent agents and patent litigators as they garner experience in their craft while also exploring the boundaries of China's Patent Law. These efforts also help to increase the sophistication of the PRB and the court system and will eventually build a truly meaningful body of legal precedent.

In fact, many of these increased efforts are already taking place. While Western commentators have been slow to appreciate the progress made in China's patent enforcement system, IP actors with a presence in China have not been. The dramatic increase in Chinese patent filings (see Chapter 3, Section 5) coupled with the increase in patent lawsuits brought in China (see Table 5.6) would certainly seem to indicate a rapidly growing confidence in China's ability to protect patent rights.

6. China's Bayh–Dole system

> For governments, granting [public research organizations] rights to IP generated with public funds can lead to better use of research results that might otherwise remain unexploited as well as to the creation of academic spin-offs or start-ups that create employment.   — OECD[1]

There should be little debate that China possesses a reasonable IP protection system that provides sufficient patent protection, including patent enforcement, to support a Bayh–Dole approach to technology transfer. The next key question, therefore, revolves around the ownership of IP generated with government funds. Under Chinese law, who owns the IP that results from government-funded research? This chapter explores China's legal framework for commercializing university-developed technology, including the economic incentives that result from that framework. This chapter also analyzes the sharp increase in technology commercialization that has come from those efforts.

1.   CHINA'S BAYH–DOLE

1.1   Earliest Regulations for IP Ownership of Government-Funded R&D

Until 1994, China did not have a well-formed policy regarding IP generated from government-funded research.[2] During the earliest years of the reform era, there was little reason to worry about such IP ownership, since the government owned everything and China's innovation system was still operating under the 'functional specialization' model (see Chapter 2). GRIs generated R&D, which the government owned and could transfer to the business sector as it saw fit. With China's attempt to create a technology market in 1985 (see Chapter 2), however, the need to incentivize and legitimize China's various technology actors (including GRIs and universities) to transfer government-funded research became relevant. In 1985, China's State Council issued Provisional Regulations of the State Council on Technology Transfer[3] (the Provisional Technology Transfer Regulations) that expressly gave Chinese universities, as well as GRIs, the right to manage, transfer and generate income from the work-related inventions of their researchers.[4] Title to the inventions officially remained with the national government (which was consistent with China's overall approach to property rights in the 1980s), but universities were given valuable 'income' rights in the IP. Specifically, Section 4 of the Provisional Technology Transfer Regulations stated:

> Technology used in research and development under the plans of the State or those of a relevant high level authority may, in addition to being disseminated for application in accordance with the provisions of the plan, be transferred in accordance with these Regulations by the unit responsible for the technology. Income from such transfers shall belong to that unit. Personnel directly involved in the research and development of the technology in question shall be rewarded. . . . Personnel who contribute by assisting the transferee in a technology transfer to fully grasp the transferred technology shall be rewarded in the same way as those personnel who directly participate in the research and development of the technology.[5]

As we explained in Chapter 2, China's initial technology market effort failed for a variety of reasons. Without a properly functioning technology market, it is not surprising that few universities asserted such technology transfer rights under the Provisional Technology Transfer Regulations.[6] Some government-funded research did reach the business sector during this period, but it tended to do so outside the Provisional Technology Transfer Regulations. GRIs were the primary recipient of government-funding for research at that time (they received 61% of China's R&D funding in 1987, compared to 35% for enterprises and 4% for universities),[7] and the various reform efforts actually caused them to be more insular. Rather than transfer technology to the business sector through licensing arrangements, GRIs were much more likely to retain their most valuable technology and try to commercially exploit it themselves.[8]

With the launch of the Torch Program in 1988 (see Chapter 2), the start-up model for technology transfer became the dominant technology transfer technique. It does not appear, however, that IP ownership concerns, or the detailed provisions of technology transfer laws, were terribly important to the entrepreneurs forming start-ups under the Torch Program. Such an attitude should not at all be surprising. During the late 1980s in China, property ownership as a whole remained a poorly defined concept and the legal system did not play a major role in economic affairs. Entrepreneurs seeking to launch Torch Program start-ups still required sufficient confidence that they would be able to capture the profits from these ventures – e.g., that the government would not confiscate their ventures or the profits generated by the ventures. Entrepreneurs were just not overly reliant on IP laws (or the legal system generally) to provide that confidence.

A possible alternative source for the necessary confidence was China's general reform environment in the 1980s. Remember, China was not a legalistic society at all in the 1980s. The law might not have been driving the entrepreneurs' decision-making at the time, but the general direction of reform movements probably was. Entrepreneurs witnessing the increasing liberalization of the economy and the government's emphasis on science and technology likely surmised that it would be okay to launch ventures with government-funded R&D . . . and these entrepreneurs were largely correct in their assumptions.[9]

1.2 Current Legislation and Regulatory Framework for Technology Transfer

As the role of law in economic affairs grew during the 1990s, the need to clarify IP ownership for government-funded R&D also increased. China responded with a series of laws and regulations throughout the decade that addressed technology transfer, including:

- Law on Science and Technology Progress (1993, as amended in 2007)
- Law on Popularization of Science and Technology (2002)
- Law on Promoting the Transformation of Science and Technological Achievements (1996)
- The Regulations on National Awards for Science and Technology (1999, as amended in 2003)[10]
- Detailed Rules for the Implementation of the Regulation on National Awards for Science and Technology (1999, as amended in 2004 and 2008)[11]
- Provisions on Promoting the Application of Scientific and Technological Achievements (1999)[12]

Overall, these laws and regulations have sought to promote S&T development in China and increase productive links between research universities and the business sector. Like the U.S. Bayh–Dole Act, China's default rule places patenting rights and IP ownership in the hands of universities. More specifically, China's legal and regulatory regime for university technology transfer provides:[13]

- Universities are expressly permitted to patent and own the IP rights that result from government-funded research,[14] but must compensate the individual inventors. The compensation requirement stems from the Chinese Patent Law requirement that an employer pay reasonable remuneration to an employee/inventor for the transfer, exploitation and licensing of an invention for hire.[15] *See* Chapter 5, Box 5.2 for a summary of the regulatory guidelines for reasonable remuneration.

- A university may decide to commercially exploit the IP itself, or it may assign or license (on an exclusive or non-exclusive basis) the IP to a third party to exploit.[16]
- The government retains certain rights in IP that result from government-funded research. The government retains a nonexclusive, royalty-free license to practice any inventions resulting from government funding. The government may decide that title to the IP should rest with the government in certain compelling cases, such as national security, vital interests of the state or compelling public policy reasons. Finally, the government possesses 'march-in' rights that allow it, under certain circumstances, to require the university to license the IP to a third party.[17]
- If the university is unable to commercialize the IP, the inventor is permitted certain rights to commercially employ the IP.[18]
- Universities establishing a start-up company may use R&D findings as an equity investment in the start-up.[19]
- University researchers are allowed to take part-time positions in business firms, so long as they are able to satisfy the responsibilities they owe to their universities.[20]
- University researchers are encouraged to form start-ups. University researchers are permitted to take a two-year leave of absence to establish a start-up. If the start-up fails, the university researcher has the right to return to her university at her prior status.[21]

It is interesting to note that China's technology transfer regulations focus on employee rights in addition to the universities' IP ownership. The increase in professor/inventor rights is a significant change from the 1980s, when university researchers were afforded very limited freedom. Chinese researcher Ming-Feng Tang explains:

In 1980s, having a part-time job outside campus was prohibited. University staffs should focus on training students. Teaching staffs that violate the rule would be punished. For example, Yu Min-hong, the founder of New Oriental (a very famous foreign language training school in China), had such a terrible experience. Being a teacher in Beijing University, he took over a class for an absent teacher outside campus. He wanted to earn additional money to support his family. The consequence was that he suffered the humiliation of being forced to resign because of campus conservative attitude at this 'derailment behavior' at that time.[22]

Since the 1980s, the attitude regarding university researchers forming productive relationships with industry has changed 180 degrees. Tang goes on to explain:

> University staffs go out of campus to serve enterprises as a part-time employee or even leave university and jump into enterprises. Having a part-time job outside campus is usually viewed as a symbol of holding a strong competence. As researchers are no longer confined in campus, they are active to serve the society by acting as problem solvers for enterprises or academic entrepreneurs. Hence, it is said that now universities have triple duties: teaching, doing researches and commercializing S&T research results (contribute to the economic development).[23]

2. TECHNOLOGY COMMERCIALIZATION RESULTS

Chinese universities have clearly responded to the government's Bayh–Dole efforts. With greater economic incentives to create and diffuse commercial technology, universities have greatly increased their technology commercialization activities. Since the late 1990s, the patent activity for China's universities has dramatically increased in each of the major categories for tracking patenting behavior. Chinese universities are applying for more patents and receiving more patent grants. They are generating more revenue from those patents by conducting more (and more lucrative) licensing and assignment transactions. Finally, university start-ups have become more vibrant, high-technology companies than ever before.

2.1 Substantial Increase in Patent Applications

China launched its first technology market effort in 1985, which was the year after China enacted its Patent Law. A number of universities responded to this regulatory change by attempting to pursue a patent strategy. In 1985, Chinese universities filed roughly 1000 patent applications.[24] As we explained earlier, however, the technology market did not take off and university patent applications remained modest through 1998.

In 1999, Chinese universities did respond to the increased regulatory and policy focus on commercializing their inventions and university patent applications began to increase rapidly (see Figure 6.1). From 1999 to 2006, Chinese patent applications filed by Chinese universities increased tenfold, from roughly 2000 applications in 1999 to almost 23 000 in 2006.

Similar to the criticism of China's general proliferation of patent applications, many contend that the increase in university patent applications

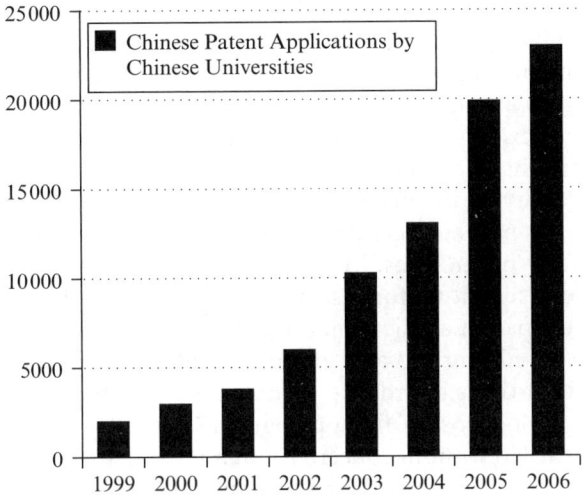

Sources: For the 2000–2006 statistics, China SIPO, available at www.sipo.gov.cn/sipo_ English/statistics/200804/t20080416_380893.html. For the 1999 figure, Enying Zheng & Hongxing Yang, *Institutions of Technology Transfer in Chinese Universities*, paper presented at the annual meeting of the American Sociological Association (Aug. 11, 2007), 5, available at www.allacademic.com/meta/p_apa_research_citation/1/8/4/9/1/1184918_index.html.

Figure 6.1   Total Chinese patent applications filed by Chinese universities (1999–2006)

has been about quantity much more than quality. For example, the 'double counting' problem that we discussed in Chapter 3 could also apply to university patent applications. Wei Hong describes the tendency of universities to apply for invention patents and utility model patents on the same invention:

> While invention patents go through a careful review process of many years, utility patents are virtually granted upon application, although a processing period of 1–2 years is inevitable. Inventors are therefore encouraged to apply for both an invention patent and a utility patent for one invention to shorten the pending period. One professor agreed with this strategy stating, 'Because it takes so long to get an invention patent, it's better to apply for a utility patent at the same time. Then my invention could be under legal protection as soon as possible.' However, another professor said, 'I myself am confused about the number of patents I have, because the TTO always asks me to apply for two patents for the same invention. They just want a higher number to show off.'[25]

Although SIPO will not issue 'double patents' – i.e., an inventor cannot obtain a utility patent and an invention patent on the same invention[26]

– it is not clear whether SIPO is able to filter out the double patent filing strategy from its 'application statistics.' As a result, it could be that patent application figures have been falsely inflated by double-counting some of the patent applications.

Another criticism lodged by some is that various government and university policies have encouraged increased patent applications, irrespective of the quality of the inventions. For example, the number of patent applications has become a crucial factor in determining university reputation and faculty promotions.[27] For some universities, patent applications have become a substitute for research publications (although the hurdle for obtaining a patent application is much lower than publishing research in a peer-reviewed journal) and graduate students may even be required to file patents at times in order to graduate.[28] One more thing to consider is the relatively low cost of filing patents in China. If patent applications are being rewarded without concern for quality and are relatively inexpensive to file, universities could be responding in a very rational manner by engaging in strategies to increase application numbers, even if the quality of the inventions does not warrant patenting. For example, universities and researchers engaging in this strategic behavior could employ a strategy of inflating patent application statistics by filing a preponderance of utility model or design patent applications.

While each of these criticisms is reasonable and even has an intuitive appeal to it, a closer look at the patent application data suggests that such critiques have been way overblown. Table 6.1 provides Chinese patent application statistics for the Chinese universities that filed the most patent applications from 1985 through 2007 (the Top University Filers). An analysis of the patent-filing activity of the Top University Filers strongly suggests that such egregious strategies for inflating patent applications are not driving the increase in patent applications.

As a group, the Top University Filers – which are the dominant filers of university patent applications – have not behaved like the overall pool of Chinese patent application filers who have historically focused on utility model and design patent applications (see Chapter 3, Section 4). The Top University Filers, by contrast, have focused on invention patent applications (73.5%) and have almost completely ignored the lowest-value design patent applications (3.8%). Rather than resemble the general pool of Chinese inventors, the application behavior of the Top University Filers much more closely resembles the behavior of foreign inventors – who are typically credited with employing more valuable patent strategies in China than domestic inventors.

Table 6.1   *Chinese patent applications statistics for Chinese universities that filed the most patent applications*

	For the period from 1985 through 2007						
	Total applications	**Invention**	**% of total**	**Utility model**	**% of total**	**Design**	**% of total**
1. Zhejiang University	7987	5849	73.2	2079	26.0	59	0.7
2. Tsinghua University	7469	6148	82.3	1315	17.6	6	–
3. Shanghai Jiaotong University	5848	5327	91.1	510	8.7	11	0.2
4. Fudan University	2927	2589	88.5	329	11.2	9	0.3
5. Tianjin University	2796	2170	77.6	613	21.9	13	0.5
6. Southeast University	2764	1526	55.2	867	31.4	371	13.4
7. South China University of Technology	2526	1703	67.4	821	32.5	2	0.1
8. Harbin Institute of Technology	2370	2001	84.4	363	15.3	6	0.3
9. Xi'an Jiaotong University	2043	1336	65.4	605	29.6	102	5.0
10. Sichuan University	1974	1654	83.8	316	16.0	4	0.2
11. Wuhan University	1938	1341	69.2	508	26.2	89	4.6
12. Jiangnan University	1837	812	44.2	68	3.7	957	52.1
13. Huazhong University of Science & Technology	1832	1172	64.0	608	33.2	52	2.8
14. Nanjing University	1660	1492	89.9	167	10.1	1	0.1
15. Beijing University of Aeronautics & Astronautics	1656	1432	86.5	222	13.4	2	0.1
16. Tongji Univesity	1653	1179	71.3	448	27.1	26	1.6
17. Peking University	1593	1480	92.9	112	7.0	1	0.1
18. Jilin University	1581	1087	68.8	491	31.1	3	0.2
19. Sun Yat-Sen University	1550	1318	85.0	220	14.2	12	0.8

Table 6.1   (continued)

	For the period from 1985 through 2007						
	Total applications	Invention	% of total	Utility model	% of total	Design	% of total
20. Shangdong University	1534	1059	69.0	472	30.8	3	0.2
21. Donghua University	1523	1140	74.9	264	17.3	119	7.8
22. Beijing University of Technology	1491	901	60.4	586	39.3	4	0.3
23. East China University of Science & Technology	1488	1328	89.3	155	10.4	5	0.3
24. Zhejiang University of Technology	1469	894	60.9	370	25.2	205	14.0
25. Shanghai University of Technology	1422	1153	81.1	242	17.0	27	1.9
Top 100 university application filers combined	105 881	77 859	73.5	24 002	22.7	4020	3.8

Source: China Education and Research Network (citing to Ministry of Education S&T Development Center), available at www.edu.cn/shu_ju_pai_hang_1088/20090622/t20090622_385824.shtml.

2.2   Substantial Increase in Patents Granted

Similar to the patent application increase, patents granted to Chinese universities has recently skyrocketed (see Figure 6.2). A more detailed examination of the statistics for patents granted to Chinese universities provides some insight into whether Chinese universities are primarily engaging in a disingenuous strategy to inflate application numbers or honestly attempting to obtain valuable property rights on their technological innovations. As Figure 6.3 shows, from 1985 through 2002, utility model patents dominated university patent grants from 1985 through 2002 – which lends support to the 'quantity over quality' arguments. During that period, 12 217 of the 17 923 Chinese patents, or 68.2%, granted to Chinese universities were utility model patents. That dominance ended in 2003, however, when the upsurge in university patent activity truly accelerated.

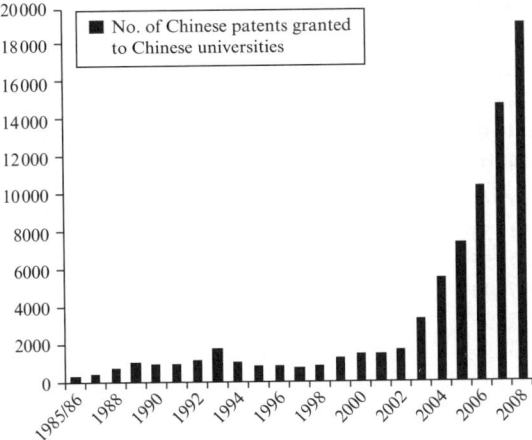

Source: China Education and Research Network (citing to Ministry of Education S&T Development Center), available at www.edu.cn/shu_ju_pai_hang_1088/20090622/ t20090622_385835.shtml.

Figure 6.2   *Total Chinese patents granted to Chinese universities (1985–2008)*

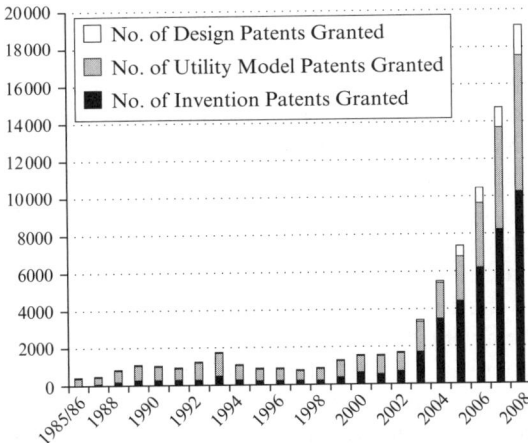

Source: China Education and Research Network (citing to Ministry of Education S&T Development Center), available at www.edu.cn/shu_ju_pai_hang_1088/20090622/ t20090622_385835.shtml.

Figure 6.3   *Chinese patents granted to Chinese universities based on type of patent (1985–2008)*

Table 6.2   Detailed breakdown of Chinese patents granted to Chinese universities (2003–2008)

	Total patents granted	Invention patents	% of total	Utility model patents	% of total	Design patents	% of total
2008	19 159	10 265	53.4	7242	37.8	1652	16.1
2007	14 773	8214	55.6	5502	37.2	1057	7.2
2006	10 457	6198	59.3	3453	33.0	806	7.7
2005	7399	4453	60.2	2391	32.3	555	7.5
2004	5505	3484	63.3	1910	34.7	111	2.0
2003	3416	1730	50.6	1582	46.3	104	3.0
2003–2008	60 709	34 344	56.6	22 080	36.4	4285	7.1

Source: China Education and Research Network (citing to Ministry of Education S&T Development Center), available at www.edu.cn/shu_ju_pai_hang_1088/20090622/t20090622_385835.shtml.

From 2003 to 2008, invention patents accounted for more than half of the patents granted to universities in each year (see Table 6.2). Universities appear to have shifted from a focus on lower quality patent protection to a higher-protection 'invention patent' strategy.

As a result of this continuously increasing patent activity, a number of Chinese universities have been able to generate significant patent portfolios. Table 6.3 provides a look at the effective patents for the 25 Chinese universities with the largest patent portfolios, together with their patent awards for 2008. While there are more than a few universities with curious patenting profiles that could indicate a strategic effort to inflate their patent statistics with lower-quality patents – e.g., their patent portfolios or 2008 patent awards are very heavily weighted toward design patents or utility model patents – the focus on invention patents is undeniable. For the top 100 universities based on effective patents, more than two-thirds of their patents are invention patents. If one looks specifically at the universities that are generally considered to be China's top research universities, the concentration of invention patents is even stronger: e.g., Tsinghua University (86.0% of effective patents are invention patents); Peking University (96.2%); Zhejiang University (74.6%); Shanghai Jiaotong University (95.7%); Nanjing University[29] (95.8%); Fudan University (84.5%); Sun Yat-Sen University (87.4%); and Xi'an Jiaotong University (86.7%).

Table 6.3   Chinese universities with the most effective patents

	Patents Granted to University in 2008				Effective Patents as of 2008			
	Total	Inv.	Util. Mod.	Design	Total	Inv.	Util. Mod.	Design
1. Zhejiang University	1023	748	267	8	2345	1750	583	12
2. Tsinghua University	679	613	66	0	2300	1977	318	5
3. Shanghai Jiaotong University	618	577	39	2	2066	1978	85	3
4. Southeast University	543	181	181	181	995	488	350	157
5. South China University of Technology	338	189	149	0	857	470	386	1
6. Xi'an Jiaotong University	228	178	23	27	709	615	71	23
7. Harbin Institute of Technology	311	287	24	0	699	635	64	0
8. Fudan University	232	188	42	2	664	561	93	10
9. Beijing University of Technology	353	144	207	2	640	268	368	4
10. Zhejiang University of Technology	383	137	109	137	630	259	200	171
11. Wuhan University of Technology	195	158	36	1	596	459	135	2
12. Donghua University	211	95	79	37	558	322	177	59
13. Tianjin University	267	241	26	0	539	452	86	1
14. Tongji University	232	154	78	0	527	330	196	1
15. Sichuan University	196	167	29	0	522	455	66	1
16. Huazhong University of Science & Technology	245	128	115	2	492	288	200	4
17. Shangdong University	231	120	110	1	467	270	196	1
18. Beijing University of Science & Technology	136	116	20	0	464	380	84	0
19. Jiangnan University	257	94	16	147	452	171	21	260
20. Shanghai University	185	140	43	2	438	339	95	4
21. Beijing University of Aeronautics & Astronautics	277	221	56	0	430	336	94	0
22. East China University of Science and Technology	103	83	20	0	428	359	69	0
23. Peking University	140	137	3	0	423	407	15	1
24. Sun Yat-Sen University	153	121	32	0	421	368	46	7

Table 6.3   (continued)

	Patents Granted to University in 2008				Effective Patents as of 2008			
	Total	Inv.	Util. Mod.	Design	Total	Inv.	Util. Mod.	Design
25. Wuhan University	236	126	80	30	414	236	143	35
Top 100 universities combined	13 997	8583	4056	1358	31 813	21 410	8487	1916
On a percentage basis	100%	61.3%	29.0%	9.7%	100%	67.3%	26.7%	6.0%

Source: China Education and Research Network (citing to Ministry of Education S&T Development Center), www.edu.cn/shu_ju_pai_hang_1088/20090622/t20090622_385829. shtml.

2.3   Commercial Diffusion of Patented Technology

Obtaining patents is only one part of an effective Bayh–Dole strategy. In addition to obtaining defensible property rights in their technology, universities also need to diffuse that technology into the market place so that it can become part of useful products and services. There are two fundamental strategies that universities employ to commercially diffuse their patented technology: (1) patent licensing or assignments; or (2) forming start-ups.

2.3.1   Patent licensing and assignments
The traditional technology commercialization strategy for universities in the West is to license, or assign, their patents. A licensing strategy requires the university to decide whether to transfer the patented technology to a single company – through an exclusive license or patent assignment – or to permit multiple companies access to the technology through non-exclusive licenses. China's technology transfer regime permits Chinese universities this same choice.

In the United States, this decision on whether to license university patents to a single company or to multiple companies tends to be guided by two major factors:[30] (1) which type of license is likely to promote the most rapid commercialization?; and (2) which type of license will best serve the public's interest? The University of California's Office of Technology Transfer explains the concept as follows:

> Patents which are broad in scope and can be used in multiple industries, or patents that they are so basic that they form the building blocks for new

technologies are most likely to be licensed non-exclusively, or by fields of use. An exclusive, 'field-of-use' license is a way to protect a market for a company while enabling the university to identify more than one license to assure public utilization of the technology in all markets.

Universities most frequently will grant exclusive licenses to patents that require significant private investment to reach the marketplace or are so embryonic that exclusivity is necessary to induce the investment needed to determine utility. Frequently, these are new drugs requiring time-intensive and capital-intensive development or they are technologies that have only a tenuous link between the workbench and production. As such, they require a company willing to dedicate financial backing and the creativity of its own scientists on a long-range basis.[31]

Chinese law provides little guidance on this single v. multiple licensee question, and it appears that the decision is basically left in the hands of the university licensor. Based on conversations with Chinese TTO officials, Chinese universities prefer to enter into non-exclusive licenses. Unfortunately, we have not been able to find any statistics that provide a detailed breakdown between exclusive and non-exclusive licenses by Chinese universities, so we are not able to verify the accuracy of the stated preference for non-exclusive licenses.

Overall, licensing/assignment revenue remains modest for Chinese universities. The most recent statistics that we were able to obtain are from 2005, which are set forth in Table 6.4. The 2005 patent license/assignment data are a few years old, so they should be viewed cautiously. Since 2005, Chinese universities have dramatically increased their patent activity (see Figure 6.3), which may have significantly increased the overall level of patent licenses/assignments. Even with that qualifier, the 2005 data are, nevertheless, striking. As of 2005, meaningful commercialization through patent licenses/assignments was restricted to only a few universities. The top four universities in Table 6.4 accounted for 63% of the revenue generated by the top 25 universities and the top eight accounted for 79% of the revenue. Once you get to the 20th ranked university on the list, the level of patent license/assignment revenues is truly immaterial.

The low revenues are not all that surprising, as it takes a long time to develop strong commercial licensing/assignment capabilities. There are a number of concerns, however, about the extremely low revenues shown by Table 6.4. Has the paucity of revenues shown in Table 6.4 continued? Assuming that license/assignment revenue has remained relatively modest which we think is a reasonable assumption, what is the likely cause? Based on our review of China's overall innovation system, it does not appear that weak license/assignment results would be due to the larger problems that derailed China's technology transfer attempts of the past. China has

Table 6.4   Chinese universities directly under the MOE – Top 25 based on 2005 patent license/assignment revenues

Rank	University	2005 revenue from patent licenses/ assignments	No. of patent licenses/assignments in 2005
1	Tsinghua University	RMB 28.00 million	70
2	Univ. of Science and Tech. of China	RMB 20.27 million	2
3	China Univ. of Petroleum – Beijing	RMB 18.77 million	48
4	Zhejiang University	RMB 11.86 million	59
5	Shanghai Jiaotang University	RMB 6.28 million	10
6	Southeast University	RMB 5.06 million	2
7	Tongji University	RMB 4.07 million	18
8	Nanjing Agricultural University	RMB 4.06 million	2
9	East China Univ. of Science & Tech.	RMB 2.86 million	10
10	South China Univ. of Technology	RMB 2.80 million	4
11	Peking University	RMB 2.69 million	3
12	Huazhong Univ. of Science & Tech.	RMB 2.63 million	12
13	Sichuan University	RMB 2.49 million	11
14	University of Electronics Science & Tech.	RMB 2.00 million	1
15	Donghua University	RMB 1.77 million	13
16	Wuhan University	RMB 1.60 million	1
17	Nanjing University	RMB 1.55 million	14
18	Shandong University	RMB 1.40 million	11
19	Dalian University of Technology	RMB 1.25 million	8
20	Xi'an Jiaotang University	RMB 0.91 million	9
21	Hefei University of Technology	RMB 0.89 million	5
22	China University of Mining & Tech.	RMB 0.50 million	0
23	Wuhan University of Technology	RMB 0.48 million	6
24	Ocean University of China	RMB 0.26 million	2
25	Beijing University of Posts & Telecommunications	RMB 0.20 million	19
Total for the Top 25 universities		RMB 124.65 million	340

Source: Ministry of Education Science & Technology (data issued Nov. 23, 2007), available at www.cutech.edu.cn/cn/dxph/qt/webinfo/2007/11/1194500769155527.htm.

developed a market-based innovation environment that is more than adequate to support university–industry technology transfer. In addition, China's legal system – including its IP law system and its laws and regulations for university technology transfer – has also developed into a sufficiently robust institution to generally support the various types of transactions and decisions that make up a vibrant Bayh–Dole system.

If China's commitment to a market-based innovation system and legal institutions that support such a system are not the problem, then what is? We believe the most likely culprits are the very same mundane problems that plague many of the weaker technology-transfer universities in the United States:

- *Local business sector is not capable of absorbing the technology:* patents tend to focus on incremental advancements to existing technologies, which requires that the potential recipient of the patent already possess expertise in that existing technology. For many Chinese companies, their R&D capabilities are at a lower level where they need to acquire 'a complete product or process that could be actually used by industry'[32] for the technology transfer to be truly useful.
- *Hard to create a technology transfer culture at research universities:* at a general level, there is always the risk that university researchers will view commercialization efforts as interfering with more traditional concepts on the role of universities. At a more specific level, technology commercialization can actually be quite taxing on the time and energy of university researchers. Most university inventions tend to be embryonic, which leads to a few serious problems. First, such inventions frequently require significant post-invention faculty participation (e.g., doing development work with licensees) to become commercially valuable.[33] Researchers that prefer basic research and publications may not be willing to devote the extra time needed for applied R&D and managing the licensee relationship.[34] Second, researchers who are not trained in the commercial aspects of inventions may not see the commercial potential of their inventions and simply fail to disclose their inventions to the university.[35]
- *A mismatch between university research capacity and the technology needs of industry:* there is no universal law that says that universities will automatically develop research that is synergistic with the technology needs and capacities of industry. As linkages strengthen between research universities and the business sector, information can flow between these groups to allow research universities to become more responsive to the needs of the business sector, while

also allowing the business sector to develop specific business models to take advantage of the technology strength of particular research universities. Where the linkages are weak, a mismatch is much more likely to occur.

- *Weak IP management:* it takes years and a large number of transactions for universities to develop truly sophisticated, efficient IP management capacity. It could be that Chinese universities simply need more time and experience to develop that IP management capacity.

It is reasonable to expect that each of these problems will tend to improve over time as more technology transfer transactions are done and each of the various actors in the Bayh–Dole systems develops valuable experience and know-how regarding university technology commercialization and the economic benefits that it can generate. We feel compelled, however, to note a major potential concern. If low revenues from patent license/assignment activity persist for too long – even if incremental progress is being made – some universities and their professors could lose faith in the viability of university technology commercialization. It can be thought of as a 'negative demonstration effect,' whereby viewing others' failures dissuades individuals from acting. Managing expectations, which we discuss in detail in Chapter 9, is very important to minimizing the impact of this 'negative demonstration effect.'

2.3.2   University start-ups

In addition to the traditional patent licensing/assignment strategy, universities can also seek to commercialize their patented inventions by creating start-ups. In the United States, universities tend to license their technology to a broad spectrum of licensees, including both for-profit and non-profit entities (e.g., other universities).[36] Regarding 'for-profit' licensees, they can range from the largest multinational corporations to very small, early-stage start-ups that are formed for the specific purpose of commercializing the university's patented technology.[37] When deciding on a licensee, U.S. universities are again guided by the overarching principle of finding the most effective means for transferring the technology so that it can benefit the public.[38] Sometimes, such an ideal licensee does not exist, in which case the university may decide to commercialize the technology itself by forming a university-owned start-up (USU).

The USU model for technology commercialization can provide some significant benefits to the commercialization process, but also tends to present a number challenges and conflicts of interest that need to be carefully managed.[39] On the positive side, a USU can lead to the most complete form of technology transfer. Frequently, the patent only covers part of the overall technology needed to create a useful commercial product or service. There is accompanying 'know-how' that the university researchers will possess that is needed to maximize the value of the patented technology. In the typical technology-based USU, the university researchers that developed the patented technology will become employees, or owners, of the USU and thereby bring their know-how with them to the USU. Moreover, by personally involving the inventor with the USU, the venture is assured of a highly-motivated entrepreneur to push for the successful commercialization of the technology – which focus can be lost when the technology is transferred to a larger company. On the negative side, technology-based start-ups have a very high failure rate – and that failure is frequently due to factors not involving the quality of the company's technology. Managerial incompetence or an inability to raise sufficient financial resources are just as likely reasons for a start-up to fail, which means that valuable technology may be trapped in a failing company. In addition, the involvement of university researchers in USUs can lead to significant conflicts of interest that need to be managed carefully.

Like their U.S. counterparts, Chinese universities frequently employ the USU strategy to commercialize patented technology. Before considering available statistics on Chinese USUs, an important explanation is in order. Chinese universities have been creating USUs since the early 1980s. Most of the early USUs, however, were not the high-tech start-up that one thinks of being associated with research universities in the West. These early Chinese USUs were predominantly non-high-tech companies that focused on such endeavors as 'producing consumer goods and providing general services (e.g. food, printing, etc.),'[40] and most had nothing to do with a technology commercialization strategy. Even today, it does not appear that Chinese universities limit their creation of USUs solely to the patent-based S&T USUs that are employed by U.S. universities as a technology commercialization strategy (i.e., the core technology for the USU originated from the university's research). As a result, statistics on Chinese USUs need to be interpreted with great care. Current statistics on Chinese USUs try to provide some guidance on the issue by categorizing USUs as being 'S&T' or 'non-S&T' USUs (see Table 6.5). While somewhat helpful, we are concerned that the S&T USU category may not be all that stringent in what is classified as S&T USUs. The S&T USU category could include a significant number of lower-technology USUs that are not truly involved with commercializing faculty research. With all that as an introduction, Table 6.5 and Figure 6.4 provide dated, but somewhat suggestive, data about Chinese S&T USUs and their relevance.

Beginning in the late 1990s, Chinese universities began to divest themselves of non-S&T USUs.[41] From 2000 to 2004, the total number of S&T

Table 6.5   Select statistics for Chinese S&T USUs

	2000	**2001**	**2002**	**2003**	**2004**
Number of S&T USUs	2097	1993	2216	2447	2355
As % of total USUs	38.5%	39.6%	43.9%	50.6%	51.6%
Profits of S&T USUs (billions of RMB)	2.8	2.4	1.9	1.5	2.4

Sources: Weiping Wu (2010), 'Managing and Incentivizing Research Commercialization in Chinese Universities', 35 *J. Technol. Transf.* 203, 210; and Xielin Liu and Nannan Lundin (2008), 'Toward a Market-Based Open Innovation System of China', in Reinhard Meckl (ed.), *Technology and Innovation Management: Theories, Methods and Practices from Germany and China*, p.21.

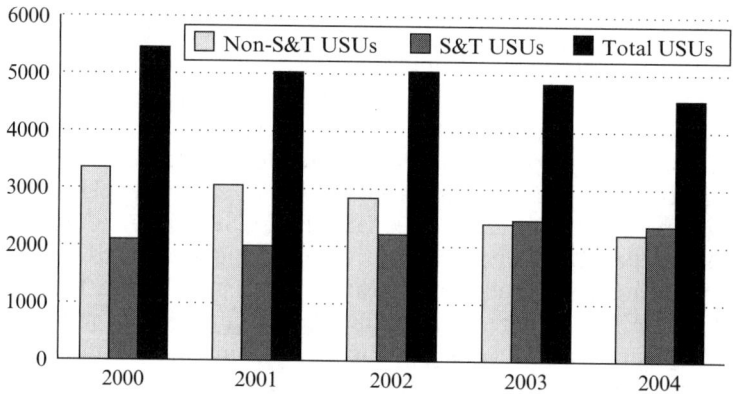

Source: Weiping Wu (2010), 'Managing and Incentivizing Research Commercialization in Chinese Universities', 35 *J. Technol. Transf.* 203, 210 and Xielin Liu & Naunan Lundin, 'Toward a Market-based Open Invitation System of China', in Reinhard Meckl (ed.), *Technology and Innovation Management: Theories Methods and Practices from Germany and China*, p.21.

Figure 6.4   Increase of S&T USUs and decrease of non-S&T USUs: 2000–2004

USUs stayed relatively stable, but the number of non-S&T USUs sharply declined as Chinese universities divested themselves of more than 1000 non-S&T USUs (see Figure 6.4).[42] Presumably, Chinese universities decided to focus on their core function of education and technology generation and divested themselves of non-essential USUs that shared little with the overall university mission.

As we explained above, the USU strategy for technology commercialization can have very real benefits, but it is not a problem-free strategy. In a recent article, Professor Weiping Wu explained some of the problems a USU strategy can cause for Chinese universities:

> In most universities, all [USUs] are under the oversight of a university enterprise group or holding company that is under the university leadership. As a result, university administration and enterprise decision-making are often intertwined. Some universities have directly invested in technology spinoffs and become the sole owners of enterprises . . . Such a relationship raises concerns about the potential financial risks that universities are exposed to . . . In [Shanghai Jiaotang University (SJTU)], for instance, the party secretary of the university serves as the chairman of the board for the enterprise group and the university president as vice chairman. The university has had to bail out some unprofitable firms from time to time, and faculty sees this as more of an intrusion to the traditional academic culture (personal interview with a SJTU administrator).[43]

Such conflicted relationships can lead to bad managerial decisions for the USUs (which impede the USUs' ability to commercialize technology) and increased financial risk for the universities. One of the more well-known examples of a university that exposed itself to an inappropriate level of financial risk by backing its own USUs is a U.S. example: Boston University. During the 1980s and 1990s, Boston University invested close to $100 million in a biotechnology company called Seragen[44] that was formed to commercialize a new cancer-curing drug developed by one of Boston University's professors.[45] John Silber, Boston University's president, and eventually chancellor, explained, 'If we had just 5 percent of what eventually became Bell Telephone, we would be richer than Harvard today.'[46] Of course, Seragen did not become Bell Telephone, and eventually went bankrupt, costing Boston University almost all of its investment in Seragen and much of the school's endowment.[47]

Chinese universities are not oblivious to the managerial and financial risk problems that come with USUs. In fact, the trend in China is for university administration to decrease their direct involvement in USU ownership, investment and involvement (see Table 6.6).[48]

While the S&T USU strategy remains a very significant technology commercialization strategy in China, we expect the relative prominence of the strategy to decline over time. As the R&D capabilities of the Chinese business sector increases and the linkages between Chinese universities and businesses increases – and we expect that both of those trends will occur – the need for Chinese universities to form their own companies to commercialize technology should correspondingly decrease. One reason for the large S&T USU sector was the lack of viable business partners to which Chinese universities could transfer technology.[49] The S&T USU strategy was not necessarily chosen as the most efficient strategy, but instead was the only viable strategy. As more technology business partners

Table 6.6   Changing ownership structure of USUs

	2000		2004	
	Number	**Percent**	**Number**	**Percent**
Total USUs	5451	100%	4563	100%
Wholly-owned by university	4793	87.9%	3044	66.7%
Joint venture with domestic partner(s)	556	10.2%	1478	32.4%
Joint venture with foreign partner(s)	102	1.9%	41	0.9%

Source: Weiping Wu (2010), 'Managing and Incentivizing Research Commercialization in Chinese Universities', 35 *J. Technol. Transf.* 203, 212.

become available, there is every reason to believe that Chinese universities will adapt to such a changing environment and shift their focus to more flexible, and lower-risk, technology commercialization strategies such as the traditional patent licensing strategy.

2.4   Technology Contracts

For all the focus that patents and patent-based strategies have received since the late 1990s, Chinese universities continue to transfer much of their commercial technology to the business sector without the use of patents. In China, businesses frequently enter into contracts for technology services (technology contracts) with Chinese universities.[50] In effect, businesses outsource some of their R&D function to a university.

When compared to the traditional patent licensing strategy, statistics from 2000 to 2004 show that technology contracts were by far the more common technology transfer technique for Chinese universities. During the five-year period, Chinese universities entered into a total of more than 33 000 technology contracts worth RMB 12.5 billion, compared to just over 2500 patent licenses/assignments worth RMB 1.3 billion (see Figures 6.5 and 6.6).

Similar to our forecasts for the USU strategy, we expect the relative prominence of the technology contract strategy to decline over time and be replaced by a greater level of patent license activity. One of the more commonly cited reasons for the popularity of the technology contract strategy has been the weak R&D capabilities of the Chinese business sector.[51] As we explained earlier, Chinese companies with weaker R&D capabilities may not be capable of doing the development work that is needed for most university patent licenses, and therefore need to acquire complete technology sets through a technology contract. As the R&D capabilities

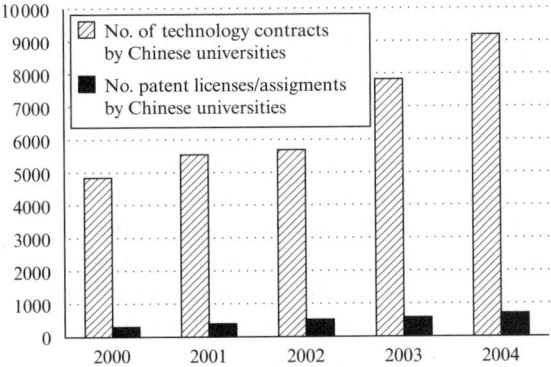

Source: Weiping Wu (2010), 'Managing and Incentivizing Research Commercialization in Chinese Universities', 35 *J. Technol. Transf.* 203, 210.

Figure 6.5 *No. of technology contracts dwarfs that of patent licenses/ assignments (2000–2004)*

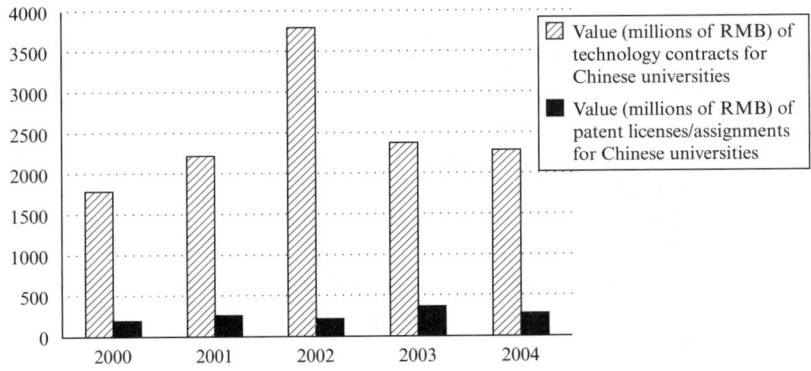

Source: Weiping Wu (2010), 'Managing and Incentivizing Research Commercialization in Chinese Universities', 35 *J. Technol. Transf.* 203, 210.

Figure 6.6 *Value of technology contracts dwarfs that of patent licenses/ assignments (2004–2004)*

of the Chinese business sector increase – and, again, we expect that will occur – the demand for patented technology from Chinese universities, rather than just technology services, should increase and the license strategy should begin to displace the technology contract strategy.

3.   DESPITE THE REMARKABLE PROGRESS, CHINESE UNIVERSITY TECHNOLOGY COMMERCIALIZATION IS ONLY IN ITS INFANCY

It will take time for university technology commercialization to play its ideal role in Chinese economic development. Despite the remarkable progress since the late 1990s, the commercialization of university inventions remains at the periphery of China's innovation system. The fact that technology contracts continue to be the dominant form of technology commercialization, while patent licenses remain on the periphery, is a clear signal that China's university technology commercialization system is still in its infancy. None of this is meant to be a criticism of the university technology commercialization system, just a statement of where it currently is in its evolution.

We fully expect the patent license strategy to become the dominant strategy. The more relevant question, from our perspective, is simply how long before that transition occurs. Generally speaking, it will occur when:

- Chinese companies have reached such an R&D sophistication level that they can conduct the development work needed to efficiently license university inventions rather than having to acquire packaged technology sets;
- Technology commercialization becomes truly integrated in the culture of Chinese universities;
- Chinese universities and businesses have developed effective communication channels that help focus university researchers on R&D that is responsive to social needs; and
- Chinese universities have developed sufficient IP management strength.

There is evidence suggesting that all of these developments are taking place and it could be that patent licenses will become the dominant commercialization strategy for Chinese universities in the not too distant future. That will be a very positive sign that China's overall innovation system has truly come of age.

7. Planning to be an innovative nation – China's national S&T plan and its impact on China's Bayh–Dole system

> [I]nnovation is the core of our national development strategy and a crucial link in enhancing the overall national strength.
> – Hu Jintao, China's President and General Secretary of the Communist Party[1]

China's universities have substantially increased their inventory of patents and their commercialization of patented technologies since the late 1990s. Two factors that will play a prominent role in whether that growth continues will be: (1) the willingness of the Chinese government to continue to fund university R&D at sufficiently high levels and the efficiency of that funding effort; and (2) the R&D proficiency of China's business sector.

One of the most critical aspects of any innovation system is whether adequate resources are being committed to R&D. Economic theory strongly suggests that for even the most free-market economies, the business sector will always under-invest in R&D. Because knowledge is both non-rival (it can be used by an infinite number of people at the same time without depriving any person of its use) and only partially excludable (it is difficult to exclude unintended parties from benefiting from ideas), knowledge tends to generate spillovers that allow unintended recipients of the new knowledge to receive its benefit without having to pay the knowledge producer.[2] In many cases, these knowledge spillovers can generate substantial social returns. For example, when a firm invests to improve one of its products or processes, it should be expected that competitors will use insights from those improvements to improve or develop their own products or services – and frequently at a much lower cost than what was required to discover the original improvement.[3] The result is that private sector R&D generates both *private returns* – returns for which the knowledge producer is compensated – and *social returns* – returns that benefit society, but for which the knowledge producer is not capable of being compensated. Numerous studies have shown that the social returns from private-firm R&D investment can be very large.[4] Because the original knowledge producer is not compensated for social returns, however, they are unlikely to serve as incentives for the original knowledge producer's R&D decisions. Consider the following example:

- Technology breakthrough X will generate private returns of $50 million to a knowledge producer and social returns of $150 million.
- The R&D cost for developing technology breakthrough X will be $100 million.

From a societal standpoint, the $100 million investment to generate $200 million of value is clearly worthwhile. From the perspective of a private knowledge producer, however, pursuing technology breakthrough X is not worthwhile as it would be a heavy loss-resulting activity. Left solely to private market forces, technology breakthrough X, which is a socially worthwhile R&D project, would not likely be pursued. Because social returns can make up an important part of the overall returns from R&D, the private sector substantially under-invests in R&D from a societal standpoint and many socially valuable discoveries are not made.[5] Government investment in research, if properly designed and executed, can help to overcome this problem. The government can act as a collectivizing agent for society as a whole, and fund R&D – frequently through research universities – that may not be justified by private returns, but that is capable of generating sizable social returns. One of the biggest reasons for the technology commercialization success of U.S. universities is the substantial and constant flow of federal funds to university R&D. Not surprisingly, university technology output is highly correlated to the level of resources that are committed to university R&D.

As was explained in Chapters 3 and 6,[6] a business sector that is equipped with strong, integrated R&D capabilities is also a prerequisite for a vibrant university technology commercialization system. Business sector R&D strength is critical to the ability of businesses to absorb technology and shapes the commercialization techniques that universities can consider. One of the reasons for the popularity in China of a lower-tech 'technology contract' strategy for university technology transfer over a 'patent license' strategy (see Chapter 6) appears to be an insufficient amount of sophisticated in-house R&D capacity in China's business sector.[7] Businesses with lower R&D sophistication are generally not capable of acquiring cutting-edge technology, but instead are limited to acquiring complete technology products or processes that are ready for immediate usage.[8] In addition to helping with technology absorption,

strong business sector R&D capabilities also permit a more efficient interface between the business sector's technology needs and the university sector's technology capabilities. Without strong in-house R&D capacity, it is
difficult for a firm's management to precisely understand its technological
needs and evaluate the strength of technology upgrade opportunities that
present themselves.

Having a clearer understanding of the Chinese government's university
R&D funding intentions and the future R&D strength of China's business sector, therefore, would be extremely helpful in projecting the success
of university technology commercialization in China going forward.
Fortunately, China provides a fair amount of guidance in the area.
China's planning tendencies from the Mao era have never been completely
abandoned. Five-year economic plans (Five-Year Plans) were long the
driving force for China's economy. As a highly 'planned economy' from
1953 to 1978, central planning – including through Five-Year Plans – was
the dominant tool used for determining how to allocate China's economic
resources and efforts. As a result of its market-oriented economic reforms,
China's practice of compulsory planning was gradually abandoned in
the 1980s, but the convention of announcing Five-Year Plans remains.[9]
Rather than establishing strict orders, China's Five-Year Plans now help
to set targets to guide economic development and to organize national
priorities and key national projects.[10] While no longer serving a core function in China's economy, Five-Year Plans do provide valuable insight into
government policy and the future direction of significant economic initiatives, including China's S&T future.

The Chinese government announced its most recent national S&T
plan in 2006 (the S&T Plan). The S&T Plan is embodied in China's 11th
Five-Year Plan (2006–10) and the accompanying Guidelines on China's
National Medium- and Long-Term S&T Development Plan (2006–20, the
S&T Guidelines). The S&T Guidelines lay out a 15-year S&T development
strategy that runs through 2020. Together, these documents establish
China's national S&T Plan with a stated goal for China to become an
'innovative nation' by 2020. The S&T Plan takes a varied approach to
improving China's S&T future. It identifies key priorities, it establishes
a series of quantitative and qualitative targets, and it outlines a series of
institutional reforms. The overarching goal, however, of all these efforts
is for indigenous technological development to become a central part of
China's future economic growth.

Table 7.1 provides a basic summary of the S&T Plan's highlights. As
Table 7.1 shows, the S&T Plan is extensive and covers a wide variety of
topics. This chapter will not attempt to cover the entirety of the S&T Plan,
but instead will focus on a few of the plan's elements that most directly

Table 7.1 Summary of China's S&T guidelines (2006–2020)[1]

Broad Economic Development Targets[2]	Upgrading Science and Technology to Meet Economic Development Targets		
	Quantitative Targets	Qualitative Targets	Improve Institutions and Incentives
Economic Growth: • Average, annual GDP growth rate of 7.5% through 2010, then 7.0% from 2011–2020 • Technological progress is to be the driving force behind this economic growth **Using Technology to Reduce Strain on Natural Environment:** • Reduce energy consumption per unit of GDP by 20% by 2010 • Reduce water consumption per unit of GDP by 30% by 2010 • Reduce emissions of major pollutants by 10% by 2010 • Increase forest cover by 1.8% by 2010	• Increase total factor productivity's contribution to China's economic growth to 60% by 2020 • Increase R&D intensity (R&D/GDP) to 2.0% by 2010 and 2.5% by 2020 • Reduce China's 'foreign technology reliance' to 30% by 2020 • Chinese inventors to be ranked in the top 5 in the world in receipt of invention-type patent grants by 2020 • Chinese authors to be ranked in the top 5 in the world in number of internationally cited scientific papers by 2020	• Improve China's capacity for indigenous innovations • Make enterprises (rather than the government) the driving force of the innovation system • Improve the higher education system • Develop a number of frontier technologies, in areas such as biology, materials technology, advanced manufacturing technology, and the information industry • Accelerate technological development in 11 major sectors (including energy and water resources) • Generate scientific breakthroughs in nanotechnology	• Encourage joint R&D efforts between enterprises and public R&D labs • Improve the financial environment for start-ups • Implement tax incentives aimed at improving domestic innovation • Employ government procurement practices that will foster demand for domestic innovation products • Expedite the implementation of a national strategy on IPRs and improve the national IPR system • Develop improved regulations for foreign technology imports and IPRs

Notes:
1. This table was inspired by a similar table produced by John Whalley and Weimin Zhou (2007), 'Technology Upgrading and China's Growth Strategy to 2020', *The Centre for International Governance Innovation Working Paper No. 21*, 5, available at www.cigionline.org.
2. The 2010 targets are from the Eleventh Five-Year Plan. The 2020 targets are from the S&T Guidelines.

Table 7.2   China's R&D intensity[1]

	1995	1998	2001	2004	2007
Gross domestic expenditures on R&D (GERD) (RMB billion)	34.9[2]	55.1	104.2	196.6	371.0
R&D intensity (GERD/GDP)	0.60%[3]	0.70%[4]	0.95%[5]	1.23%	1.49%

Sources:
1. MOST (2008), *China S&T Statistics Databook* (unless otherwise indicated by a specific footnote). MOST (2008), *China S&T Statistics Databook* is available at www.sts.org.cn/sjkl/kjtjdt/data2008/cstsm08.htp. MOST (1998–2007), *China S&T Statistics Databooks* are available at www.most.gov.cn/eng/statistics/2007/index.htm.
2. MOST (1998), *China S&T Statistics Databook.*
3. *Id.*
4. MOST (2004), *China S&T Statistics Databook.*
5. MOST (2007), *China S&T Statistics Databook.*

impact university R&D funding and the R&D capabilities of China's business sector. Specifically, this chapter will focus on the S&T Plan's call to: (1) dramatically increase China's R&D expenditures; (2) develop enterprises (rather than the government) as the driving force for China's innovation system; and (3) improve China's domestic innovation capacity and reduce its reliance on foreign technology.

1.   INCREASE R&D SPENDING

China already commits a lot of resources to its R&D effort, and the S&T Plan calls for an even greater commitment. The S&T plan calls for China to increase its R&D intensity to 2.0% by 2010 and to 2.5% by 2020. R&D intensity is calculated as the country's gross domestic expenditures on R&D (or GERD)[11] divided by its GDP. Table 7.2 provides summary data on China's R&D intensity from 1995 to 2007. Since 1995, China's R&D intensity has increased roughly 2.5 times – from 0.60% in 1995 to 1.49% in 2007. Because of the rapid annual increase in China's GDP, China's

GERD had to increase by more than tenfold in order to generate that 250% increase in R&D intensity.

China's 2.5% R&D intensity target for 2020 would place China in elite company in terms of R&D intensity. Table 7.3 provides the most recently available R&D intensity figures for a select group of countries and economies, including the most developed countries and economies.

1.1   Is China's 2020 R&D Intensity Target Realistic?

Achieving the 2020 target will require that China greatly expand its support for R&D. Is it realistic to believe that China can attain that level of R&D spending? While the 2.5% goal appears to be quite aggressive, it is not completely out of line with China's R&D spending patterns over the last two decades. Countries tend to follow a relatively consistent pattern for R&D spending based on their level of economic development. Lower-income economies do not spend very much on R&D – typically less than 1% of their GDP.[12] Presumably, such countries are so low on the technology food chain that they do not have enough technology-capable human resources or worthy R&D projects to justify spending considerable resources on R&D. Instead, such countries tend to devote minimal resources to R&D – with most of those resources going to identifying and absorbing foreign technology.[13] Middle-income economies tend to devote more resources to R&D – e.g., greater than 1% of GDP – than lower-income countries.[14] As these countries ascend the economic ladder, they find themselves capable of attracting, adapting and absorbing higher quality technology and they may even begin to conduct limited amounts of basic research.[15] Not surprisingly, developed economies spend the most on R&D, as they have the techno-capable human resources and worthy R&D projects that justify higher levels of spending. As Table 7.3 shows, an R&D intensity around 2% is not atypical for developed economies.

Based on China's level of economic development, its current R&D intensity is quite strong, but not completely surprising.[16] Of the world's low-middle-income countries, China is the only country with an R&D intensity above 1%.[17] At the same time, China is not like most low–middle-income countries – no other such country has generated three decades of close to 10% annual growth – so it is not entirely surprising to see that China has sought to encourage an R&D intensity that is higher than 'normal.' To provide some perspective to China's R&D intensity, Figure 7.1 compares China's R&D intensity to that of the other BRIC economies – Brazil, India and Russia.

Is it a positive that China is spending so much on R&D? The answer to that question is a definite 'maybe.' On the positive side, it demonstrates

Table 7.3   R&D intensity for select countries/economies – most recent year

Country/Economy	R&D Intensity	Country/Economy	R&D Intensity
Israel (2005)	4.71%	France (2005)	2.13%
Sweden (2005)	3.86%	Canada (2006)	1.95%
Finland (2006)	3.51%	Belgium (2005)	1.82%
Japan (2004)	3.18%	Netherlands (2004)	1.78%
South Korea (2005)	2.99%	EU-25 (2005)	1.77%
Switzerland (2004)	2.93%	Australia (2004)	1.77%
Iceland (2003)	2.86%	United Kingdom (2004)	1.73%
United States (2006)	2.57%	Luxembourg (2005)	1.56%
Germany (2005)	2.51%	Norway (2005)	1.51%
China's 2020 Target	2.50%	China (2007)	1.49%
Austria (2006)	2.44%	Czech Republic (2005)	1.42%
Denmark (2005)	2.44%	Ireland (2005)	1.25%
Taiwan (2004)	2.42%	Slovenia (2005)	1.22%
Singapore (2005)	2.36%	Brazil (2006)	1.06%
All OECD (2004)	2.25%	India (2006)	0.70%

Source (except for Brazil and India):   National Science Board (2008), *Science and Engineering Indicators 2008*, 4–40, available at www.nsf.gov/statistics/seind08/ [hereinafter NSB Report]. *Source for Brazil:* OECD (2008), *Science, Technology and Industry Outlook 2008* at 162 [hereinafter 2008 STI Outlook]. *Source for India: id.* at 22 (the 0.70% is an estimation based on a bar chart).

that China has a very real commitment to R&D that is being backed by the necessary resources. On the negative side, there is probably a very good reason that R&D intensity tends to grow gradually over time. In order for the R&D resources to be deployed efficiently, a massive collective effort is required. A critical mass of researchers capable of deploying the resources and generating meaningful R&D is required, which takes a considerable amount of time to develop. It also takes an extensive, and highly developed, array of institutions – e.g., financial markets, venture capital markets, legal system, intellectual property regime, corporate governance system (just to name a few) – to help direct the resources to worthy R&D projects and to help ensure that those resources are not wasted once deployed. These institutions and mechanisms also take an exceptionally long time to develop.

Not surprisingly, much of China's R&D spending over the last decade has been directed towards importing foreign technology. An importation strategy can help to avoid the problem of needing to wait for a critical

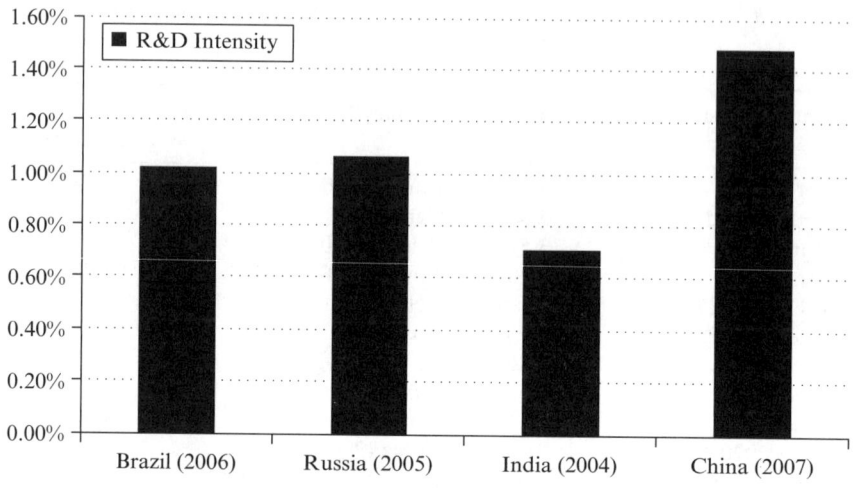

Sources: For Brazil: 2008 STI Outlook, at 164. *For India:* OECD, Country Statistical Profiles 2009: India, http://stats.oecd.org/viewhtml.aspx?queryname=18185&querytype= view&lang=en. *For Russia:* NSB Report, at 4–40. *For China:* MOST (2008), *China S&T Statistics Databook* (2008).

Figure 7.1   R&D intensity of the BRIC economies – most recent year

mass of researchers and the development of adequate institutions. The S&T Plan, however, calls for China to de-emphasize technology importation in favor of domestic R&D efforts. That means China will face the double challenge of both spending considerably more money on R&D and putting much greater stress on its relatively new institutional framework with the hope that it is capable of efficiently directing the increased resources to worthy R&D projects. This is not an impossible task, but it will be very challenging, which means there is a good chance that China's aggressive goal of 2.5% R&D intensity by 2020 will either not be met or, if it is met, will involve significant amounts of wasted resources.

1.2   How Much of the Projected Spending Increase will go to University R&D?

China's projected increase in R&D expenditures provides an overall picture of the intensity of China's future R&D efforts, but it does not provide any information about the level of resources that will be specifically allocated to the university technology commercialization effort.

How much of the increased R&D spending will be dedicated to university R&D? Not surprisingly, the S&T Plan does not provide that level of specific detail. We can, however, look at historical trends in R&D funding to universities, including the sources of those funds, for guidance on the likely distribution of the increased R&D spending.

Chapter 3 (Section 3.4) provided detailed statistics on the growth of R&D expenditures for China's higher education sector since 1998. For purposes of this trend analysis, the highlights of those statistics were:

- Government funding makes up a majority of the R&D expenditures for the higher education sector – roughly 55% each year from 2004 through 2007. As a result, the government has the greatest ability to influence the level of R&D expenditures to universities. Since increased R&D funding to universities is well-aligned with the overall policy objectives of the S&T Plan, it would seem likely that the government will be motivated to continue to increase its R&D funding for universities.
- From 1998 through 2007, R&D expenditures for the higher education sector increased roughly in line with the increases to China's overall GERD. During that period, the average annual increase for China's overall GERD was 23.7%, compared to 21.0% for higher education R&D expenditures.

While past performance is no guarantee of future results, this historical growth trend coupled with the fact that the government is the majority source for those funds certainly seems to place universities in a strong position for future R&D expenditure increases. There does not seem to be any reason to believe that China's higher education sector will not continue to experience strong R&D spending increases as China works to meet its 2.5% R&D intensity target by 2020.

2. INCREASING THE BUSINESS SECTOR'S ROLE IN THE INNOVATION SYSTEM

Committing more resources to R&D provides only a partial solution to improving an innovation system. More resources can be helpful, but they are only truly helpful if deployed intelligently. In addition to having enough resources, a successful innovation system must also encourage an efficient allocation of the resources that it commits to the R&D effort. It needs to encourage its researchers to pursue the most worthy R&D projects and ensure that those projects are the most likely to be funded.

Table 7.4   *Developed economies: business sector dominates R&D*
expenditures

	Percentage of R&D expenditures performed by			
	Business sector	**GRIs**	**Higher education sector**	**Other**
USA (2007)	71.9	10.7	13.3	4.2
Japan (2006)	77.2	8.3	12.7	1.9
Germany (2007)	70.0	13.7	16.3	–
France (2007)	63.2	16.5	19.2	1.1
UK (2006)	61.7	10.0	26.1	2.2
S. Korea (2006)	77.3	11.6	10.0	1.2
Canada (2007)	54.4	9.2	36.0	0.4
Total OECD (2006)	69.1	11.4	17.2	2.6
EU-27 (2006)	63.1	13.4	22.3	1.2

Source: OECD, *Main Science and Technology Indicators* (Vol. 2008/2).

Markets offer a variety of possible solutions to this basic allocation problem, and the S&T Plan embraces one of the most powerful of these market-based solutions with its call to further develop the business sector as the driving force for China's innovation system.

2.1   The Business Sector Plays the Dominant Role in Advanced Innovation Systems

In highly-developed countries, the business sector tends to play the dominant role in their innovation systems. In addition to being the primary mechanism for putting technology to productive use by developing and selling technology-based products and services, the business sector also accounts for the majority of the country's R&D efforts in most developed economies (see Table 7.4). The business sector's dominant role should not be surprising, as it provides a number of allocation advantages to an innovation system when compared to a predominantly government-driven system. The business sector is highly motivated to participate in productive R&D activities because it is the actor that is most directly able to capture the economic benefit from technological advances. It is the business sector that reduces technology to usable products and services and then sells those products and services. Of equal importance is the business sector's close proximity to the general public's needs and desires. As the actor that most directly profits (or loses money) based on the technology decisions it pursues, the business sector receives extremely valuable information on what technologies will be most useful to, and most desired by, the general public. The business sector must also develop discipline – based on having to answer to customers – which forces the business sector to focus resources on the most valuable technology projects and avoid waste.

As we have discussed throughout this book, China has made substantial progress in moving from a GRI-centric innovation model to an industry-centric model. In 2007, for example, China's business sector accounted for 72% of China's R&D expenditures. Despite the many successes, the overall quality of China's business sector R&D capabilities remains questionable (see Chapter 3, Section 4.2). The S&T Plan's call for enterprises to be the driving force for China's innovation system – even though it already accounts for more than 70% of R&D expenditures – appears to be an admission by China's leadership that the quality of China's business sector R&D capacity is not sufficient and that improvements are needed.

2.2   Policy Tools to Support Business Sector R&D in China

The S&T Plan employs a number of traditional policy tools to support business sector R&D in China. This section will provide a brief overview of some of the more relevant policy tools. It is far too early, however, to measure the actual impact these policy tools have had, or will have, on the R&D strength of China's business sector.

2.2.1   Tax incentives

R&D tax incentives have become a very popular policy tool among OECD countries. Rather than collect tax revenues from the business sector and entrust the government to wisely redeploy those funds on worthy R&D projects, there has been a trend toward offering greater tax relief to businesses that engage in certain types of R&D investment. In 2008, 21 OECD countries provided their businesses with tax relief for R&D investments, compared to 18 in 2004 and 12 in 1995.[18] In addition, the overall generosity of these R&D tax relief packages has tended to grow.[19]

China has been employing tax incentives to encourage business sector R&D since the 1990s, and the S&T Plan has significantly strengthened those efforts.[20] A 2008 OECD report found that China has one of the most generous R&D tax incentive programs – trailing only Spain, Mexico, and France.[21] Table 7.5 provides a brief summary of some of the more important tax incentives that have arisen from the S&T Plan.

One of the most significant features of China's newer tax incentives is their applicability to a greater pool of businesses. By increasing the pool of businesses that can benefit from R&D tax incentives, China's newer

Table 7.5   Select tax incentives that arose from China's S&T Plan

	New tax rules	Prior approach
R&D deduction	150% deduction for R&D expenses, with a five-year carry forward period.	Limited to profitable companies whose R&D expenses had increased by 10% from the previous year. The R&D deductions could not be carried forward. The prior approach discriminated against newer high-technology companies, who tend not to be profitable for years, as well as companies seeking to make substantial R&D investments that could make them unprofitable in the year of investment.
High-tech enterprises	Two-year tax holiday after profitability and reduced tax rate (15%) for State-encouraged new technology and high-technology enterprises.	Favorable tax treatment was limited to foreign-funded enterprises in high-tech parks.
Duty-free importation	Duty-free importation of certain R&D equipment for approved entities	Less generous treatment
Venture capital	Preferential tax treatment for venture capitalists that invest primarily in small- to medium-sized high-technology companies	None

Source: John Whalley and Weimin Zhou (2007), 'Technology upgrading and China's growth strategy to 2010', *The Centre for International Government Innovation Working Paper No. 21*, available at www.cigionline.org, at 14–16.

tax rules should help to spread R&D investment to a greater number of enterprises in China.

One tax issue that has received particular attention in China has been the more favorable tax treatment provided to foreign-invested firms over domestic enterprises. China has offered preferential tax treatment to foreign-invested firms since the early 1990s in order to encourage foreign direct investment (FDI) into China.[22] In March 2007, however, China's National People's Congress enacted a unified Enterprise Income Tax Law, which went into effect in 2008. One of the major purposes of this new Enterprise Income Tax Law was to eliminate the differential tax treatment between foreign-invested firms and domestic enterprises so that they could compete on a more level playing field.[23] This is another issue that will be worth careful scrutiny. At some point, China should become such an attractive destination for foreign-invested firms – e.g., due to China's favorable domestic market and lower-cost, but highly-skilled, human capital – that special tax treatment is no longer needed to encourage FDI. Has China reached that point? Maybe it has. If China has become sufficiently attractive for foreign-invested firms, the new Enterprise Income Tax Law should be beneficial to China's domestic business sector by making them more competitive with their foreign counterparts. If, however, China has not yet become sufficiently attractive, the new Enterprise Income Tax Law could discourage the arrival of new foreign-invested firms which could retard the technological growth of China's overall business sector since foreign-invested firms have been a valuable source of cutting-edge technology. The Chinese government, as well as Chinese tax policy researchers, should be closely monitoring this issue.

2.2.2 Government procurement of technology products

Procurement practices have long been used by governments to assist particular classes of businesses to develop. Procurement practices help these identified businesses by creating focused government demand for their particular products or services. In the United States, for example, government procurement practices have long provided preferential treatment to smaller businesses, as well as minority-owned businesses. Chinese procurement practices are being used to create demand for domestic technology products.[24] Because such procurement practices are covered by China's WTO accession, there are significant limits on how far this practice can be pushed.[25] It will be interesting to see if China's government takes an aggressive approach to its procurement practices, which could result in numerous challenges from WTO member-countries, or a conservative approach that will allow China to avoid such WTO troubles.

2.2.3 Financial services

Access to financing is frequently a critical issue for businesses seeking to make R&D investments. Without access to capital, businesses simply cannot make the necessary investments required for meaningful R&D. China is trying to encourage business sector R&D development by improving the financing environment. Chapter 8 will examine the Chinese government's efforts to develop a world-class venture capital industry to help finance and nurture its emerging industry of high-technology start-up firms.

3. IMPROVE CAPACITY FOR DOMESTIC INNOVATION AND REDUCE RELIANCE ON FOREIGN TECHNOLOGY

To date, China's ability to accumulate technology has been largely the result of its ability to 'import' technology, rather than generate it domestically. Beginning in the 1990s, China's popularity as a destination for FDI began to increase, and China is now one of the largest annual recipients of FDI (see Table 7.6).

While FDI has provided China with a substantial boost to its economic growth, it is interesting to note that the primary benefit of FDI has not been financial. Unlike with many developing countries, China really did not need the 'money.' This was not a situation where China's banks lacked sufficient funds to finance worthy projects.[26] Instead, encouraging FDI and foreign-invested firms has helped to serve as a major channel for China to import technology, know-how and skills from developed countries.[27] This technology importation strategy has helped China to reduce its technology gap with the developed countries and has been an important part of China's high growth rate.[28]

There are limitations, however, to an S&T strategy that is primarily based on importing technology, and recent perceptions within China are tending to view the impact of imported technology, know-how and skills 'as lower than expected.'[29] Imported technology, which comes in primarily through the business sector, has not sufficiently increased domestic innovative capacity by the importing Chinese businesses.[30] It does not appear that Chinese businesses are using imported technology as a temporary strategy that helps them to jumpstart their own technology creation. Instead, many Chinese businesses are stuck in a technology buying cycle 'of purchasing foreign technology, then purchasing again the upgraded technology when the previous version is outdated.'[31]

Frequently, FDI and imported technology come into China through foreign-invested firms – e.g., multinational corporations establishing joint ventures or wholly-owned subsidiaries in China. These foreign-invested

Table 7.6   FDI flow into China

1990–2000 (annual average)	2004	2005	2006	2007
$30.1 billion	$60.6 billion	$72.4 billion	$72.7 billion	$83.5 billion

Source: UNCTAD (2008), *World Investment Report 2008, Country Fact Sheet China*, available at www.unctad.org/sections/dite_dir/docs/wir08_fs_cn_en.pdf.

Table 7.7   China's foreign technology reliance in 2004 and 2005

	2004	**2005**
Foreign technology purchases (FTP)	US$14 billion	US$19 billion
GERD	US$24 billion	US$30 billion
Foreign technology reliance (FTP/(GERD + FTP)	37%	39%

Source: John Whalley and Weimin Zhou (2007), 'Technology Upgrading and China's Growth Strategy to 2010', *The Centre for International Governance Innovation Working Paper No. 21*, available at www.cigionline.org, at 10.

firms have shown considerable high-technology success. For example, foreign-invested firms generated 88% of China's high-technology exports in 2005.[32] There is a growing dissatisfaction in China, however, with the level of technology spillovers that result from foreign-invested firms. These foreign invested firms are conducting little R&D in China, and the foreign parties tend to control the core technology.[33] China is being excluded from much of the technology benefits generated by these foreign-invested firms.

China's leadership appears to have concluded, quite reasonably, that its technology importation strategy has 'run its course.'[34] The importation strategy has helped to increase China's overall inventory of technology assets by allowing it to absorb existing, advanced technologies. But, if China wants to fully appreciate the benefits of technological progress, it will need to become more technologically self-sufficient. First, the owners of the technology (e.g., the patent holders, rather than the licensees) tend to generate the most profits, and Chinese companies are frequently excluded from these greater profits. Second, China is not able to exert its full influence in shaping the direction of technological progress, but instead is forced to depend on technology decisions that are made abroad. China, for example, has pressing needs in areas such as 'energy, water and resource utilization, environment protection, and public health,'[35] that could benefit from focused technology efforts in China.[36]

The national S&T plan seeks to reduce China's foreign technology reliance from roughly 40% in 2004 and 2005 (see Table 7.7) to 30% by 2020. Foreign technology reliance is calculated as foreign technology purchases (FTP) divided by (GERD + FTP). China's desire to reduce foreign technology reliance has drawn some concern from the international community that China may be regressing to a 'techno-nationalist' or other imprudent self-reliance strategy from the Maoist era.[37] While we do not profess to be experts on China's motivation, it strikes us that such concerns are misguided. The Chinese government's explanation that China remains committed to participating in the global innovation system, but that China also wishes to benefit from the value that stems from developing valuable IP,[38] is a very logical explanation for its strategy. Moreover, because China's R&D expenditures project to be substantially higher in 2020 (in particular if China is able to reach its 2.5% R&D intensity target), China will still be a major technology importer in 2020 with a 30% foreign technology reliance rate. In fact, China would be a substantially larger foreign technology importer in 2020 than it is today.

4. WHAT IS THE LIKELIHOOD OF THE S&T PLAN ACTUALLY BEING CARRIED OUT?

It is one thing to develop a plan, and quite another to actually carry it out. How seriously should we take the S&T Plan? One way to evaluate the likelihood that the S&T Plan will be carried out is to analyze how critical S&T development is to China. If S&T development is truly critical, and that fact is appreciated by China's leaders, self-preservation should help to incentivize them to take the plan more seriously. Box 7.1 provides a brief description of the importance of technological progress to sustainable economic growth. All indications are that China's leadership fully understands that improving the country's S&T capacity is necessary for China to meet its GDP growth goals.

Manufacturing currently accounts for roughly half of China's GDP – with half of that total being exported.[39] While exploiting lower-cost labor to churn out massive quantities of exportable goods has helped China to grow from one of the poorest countries in the world to a very respectable level, China's leadership has long understood that simply continuing that strategy will not be sufficient for China to maintain its historically high growth rates.[40] To begin with, China's manufacturing sector has now become so large – projected to be the largest producer of manufactured goods within the next decade[41] – that it is simply unrealistic to believe that it can continue at close to 10% growth rates. It is one thing to grow 10% annually from a relatively small, immature base. It is entirely different to maintain a substantially above average economic growth rate when China's manufacturing sector is both mature and extremely large.[42] Moreover, there are indications that manufacturing costs are increasing in China, which decreases its attractiveness as a haven for outsourced manufacturing and threatens a long-term, low-cost manufacturing strategy. Basic manufacturing is largely a commodity, which means that the party that agrees to do the manufacturing the cheapest will generally win the contract, since there will be many different parties that can all do the

BOX 7.1   TECHNOLOGICAL PROGRESS IS THE
KEY TO SUSTAINABLE ECONOMIC
GROWTH

Reduced to its simplest concept, economic activity involves taking physical resources and rearranging them into things (products or services) that are more valuable than before. Because physical resources are scarce, economic growth requires more than just rearranging more physical resources, but eventually requires rearranging the physical resources better.[1] That is where techno-logical progress comes into play. By improving China's ability to generate, absorb and employ science and technology, China can generate more sustainable economic growth, and do so with less damage to the environment and fewer physical resource prob-lems (e.g., energy consumption or strained water supply). Noted Stanford growth economist Paul Romer provides the following simple cooking analogy to explain the concept:

> To create valuable final products, we mix inexpensive ingredients together according to a recipe. The cooking one can do is limited by the supply of ingredients, and most cooking in the economy produces undesirable side effects [(e.g., pollution)]. If economic growth could be achieved only by doing more and more of the same kind of cooking, we would eventually run out of raw materials and suffer from unac-ceptable levels of pollution and nuisance. Human history teaches us, however, that economic growth springs from better recipes, not just from more cooking. New recipes generally produce fewer unpleasant side effects and generate more economic value per unit of raw mate-rial.[2]

Notes

1. Paul M. Romer (2007), 'Economic Growth', *The Concise Encylopedia of Economics* (David R. Henderson, ed.).
2. *Id.*

manufacturing at a sufficient quality level. All the manufacturer has to do is follow the specifications laid down by someone else.[43] Other locations in Asia (e.g., Indonesia, Vietnam and Cambodia) could begin to challenge China on the manufacturing front, which limits the economic growth opportunities from manufacturing.[44]

Even if economic growth could somehow be sustained by continuing with an intense focus on basic manufacturing, China's experience with the basic manufacturing strategy has not been problem-free. Environmental degradation, over-consumption of natural resources, and rising inequalities (in particular between China's urban and rural populations) have been very real by-products of the country's recent growth. China's leadership has made it clear that it does not want to continue to pay the high-price involved with such a heavy, basic manufacturing strategy.[45] Xu Guanhua, China's Minister of Science and Technology, recently explained:

> Our per capita supply of energy, water and land resources is becoming more stringent and the problems of resources and environment are becoming increasingly pressing after all these years of economic development. Only by promoting scientific and technological development can China solve all these problems.[46]

China's leadership has long appreciated the need for the country to commit to a viable S&T strategy. In fact, improving China's technology capabilities has been a unifying force for much of China's economic policy for the last three decades.[47] In addition to such direct matters as government R&D spending, technology development has driven China's education policy (the need to develop China's human capital base), its import policy (facilitating foreign direct investment), its export policy (emphasis on increasing high-tech exports) and a wide variety of its legal and economic reform movements. China's emphasis on intellectual property rights – which were developed in advance of private real property rights – and its corporate governance reforms that help encourage the development of high-tech firms (by providing managers ownership interests in their firms) have been grounded in China's overall technology policy. Much of China's financial sector has been similarly influenced by technology policy, allowing foreign venture capital firms to come into China while allowing Chinese technology firms to tap into more sophisticated financial markets outside China (such as Hong Kong's financial markets) to satisfy their capital requirements.

If the Chinese government's motivation ends up being the most critical factor in whether or not the S&T Plan is diligently pursued – as opposed to factors such as economic or political crisis which could always derail the plan – there is no reason to doubt that the S&T Plan will be diligently pursued.

8. China's emerging venture capital industry

> Innovations fail to create value when they cannot attract the resources required to develop them.   – Paul A. Gompers and Josh Lerner[1]

As we have tried to highlight throughout this book, one of the biggest challenges to the future success of China's Bayh–Dole efforts will be the business sector's R&D strength.[2] In this chapter, we want to focus on one particular element of China's business sector that tends to be a very natural technology transfer partner for universities in developed countries: high-technology start-up firms. In the United States, for example, such smaller, high-growth firms are critical licensees for university-developed technology (see Table 8.1).

While such detailed statistics are not readily available in China for a

Table 8.1 *University-developed technology licensees in the United States (based on AUTM's annual U.S. licensing activity surveys)*

	2005	**2006**	**2007**	**2008**
U.S. universities that responded to survey	158	161	161	159
Commercial licenses & options:				
Total executed	4201	4192	4419	4376
Executed to small companies (excluding university-created start-ups)	2193	2127	2150	2139
% executed to small companies	52.2%	50.7%	50.0%	48.9%
Executed to university-created start-ups	586	698	764	733
% executed to university-created start-ups	13.9%	16.7%	17.8%	16.8%
Executed to large companies	1203	1327	1383	1504
% executed to large companies	28.6%	31.7%	32.2%	34.4%

Sources: The Association of University Technology Managers (AUTM), *U.S. Licensing Activity Surveys – FY 2005–FY 2008.*

direct comparison, it is clear the role played by China's start-up industry in commercializing university technology pales in comparison to that of its U.S. counterpart. The primary reason for this smaller role is the still largely nascent state of China's start-up industry. Like most countries around the world, China aspires to develop a start-up industry that can someday rival the success of the companies that congregate in U.S. start-up hotbeds such as Silicon Valley or the Route 128 corridor near Boston. In the United States, start-ups have demonstrated the ability to create a disproportionate amount of the country's macroeconomic growth,[3] innovation,[4] and net new jobs.[5] Recognizing the importance of start-ups to a vibrant, innovative economy is simple enough, but developing an environment that consistently fosters their creation and growth is a daunting task that has proven to be very challenging for most countries other than the United States.

One factor that is crucial to a successful start-up environment is specialized sources of capital to fund these high-growth, high-risk companies. The creation and survival of start-ups is highly dependent on their ability to procure capital.[6] Without sufficient funds, these companies simply cannot be built. China has been working to create a more favorable financing environment for its start-up industry since the 1990s,[7] with a particular focus on developing a modern venture capital industry that is capable of promoting the growth and creation of a vibrant start-up industry. This chapter will examine China's policy efforts to promote venture capital and consider the future of China's start-up industry.

1.   START-UPS ARE DIFFICULT TO FINANCE

1.1   Information and Management Problems Plague Start-ups

Like most sophisticated economies, China's financial markets have shown a general ability to finance large, established firms. Traditional financing sources – primarily commercial banks – are able to offer a variety of debt financing options that are relatively accessible to these large, established firms. In addition, China has been developing its public equity markets, namely the Shanghai and Shenzen Stock Markets, to provide an alternative source of capital for established firms. China's public equity markets have a long way to go when compared to established public equity markets like those in the United States or the United Kingdom, but they are developing. Moreover, many of China's stronger companies have been able to access foreign capital markets, including the New York Stock Exchange or Nasdaq, to meet their capital needs.

Whether in China, the United States, or anywhere else in the world, the financing environment for young start-ups, in particular those in the high-technology sectors, is much more challenging than it is for large, established firms. Start-ups pose a number of systematic problems for traditional financing sources that cause most to shun financing start-ups. To begin with, material debt financing is not an optimal financing technique for most start-ups until they are relatively mature companies. Because of the way that start-ups are structured, they are generally not able to repay any significant debt obligations for quite some time. Start-ups are built for growth and will generally sacrifice near-term profitability for this growth. They tend to face several years of negative earnings and, therefore, lack the necessary excess cash flow to make the required principal and interest payments.[8] To compound matters, start-ups generally do not have meaningful securable assets,[9] which is a fundamental requirement for most loans to high-risk borrowers. The primary assets for a start-up are likely to be IP assets, which are very difficult to collateralize.[10]

Equity investments, therefore, are the more logical financing option for most start-ups. Equity investments involve selling an ownership share in the company. Fundamentally, an equity investor gives money to a company in exchange for a right to that company's future profits. Unfortunately for start-ups, most traditional equity investors also refuse to finance start-ups. The reluctance of the traditional equity sources stems from three systematic problems that plague start-ups and that such traditional investors are not well equipped to manage:

- *Uncertainty about future success*: young start-ups lack a historical track record, which makes judging their future performance particularly difficult. Since young start-ups are frequently building their businesses around new technologies, new markets and unproven management teams, evaluating their future prospects is extremely challenging and requires specialized expertise.[11]
- *Informational asymmetries*: asymmetric information occurs when one party to a negotiated transaction has materially less accurate information than the other party. For young start-ups, company management has substantially better information about the potential positives and risks involved with investing in the company than do potential investors.[12] In particular, the '[d]etailed technical information (especially acute for high tech ventures), and the value and merit of the technological advancement, is usually best understood by [company management].'[13]
- *Dependence on management execution*: young start-ups are particularly dependent on their management teams to successfully run the company. While much of the focus on young start-ups revolves

around their potentially revolutionary new ideas, the success of a given start-up is usually less dependent on the quality of its ideas and much more dependent on the ability of its management team to competently execute the business plan.[14] Despite the extreme importance of management execution, the relative immaturity of most start-ups and their constrained financial resources tend to result in their having inexperienced and/or thin management teams. To compound matters, the start-up management team is likely to suffer from substantial conflicts of interest based on the classic 'agency problems' (see Box 8.1) that arise in external equity investments due to the separation of ownership from control.[15]

While all firms (including well-established firms) must address each of these problems, start-ups present particularly extreme cases. To make an informed investment decision, therefore, potential start-up investors need to address these information and management problems. Specifically, potential investors should be expected, among other things, to employ a substantial and costly screening process (which typically requires specialized knowledge of the start-up's technology and business) and to actively monitor the start-up's management team.[16] Traditional financing sources tend not to be equipped to accurately identify which start-ups are truly worth financing, nor are they equipped to provide the monitoring necessary to ensure that the start-ups' managers behave appropriately.

1.2 Venture Capital Firms have Adapted to Operate Effectively in this Difficult Investing Environment

Absent some effective mechanism to overcome these problems, the credit constraints that plague start-ups globally would be substantially worse and many fewer start-ups would likely be created and developed. Fortunately, such an effective mechanism has developed in some countries – professional venture capital firms. The modern venture capital industry originated in the United States in the mid-1940s[17] and has since grown into a highly professionalized industry that annually invests multiple tens of billions of dollars of investment capital around the world. The primary focus of this chapter is on the 'professionalized' element of the venture capital industry. The definition of a venture capitalist can be 'slippery' in the real world, as that term is not always used uniformly – in particular when viewed from a global perspective.[18] To provide some structure to this concept, we are dividing the world of venture capital investors into three commonly employed designations: professional venture capital firms; angels; and private equity firms.

BOX 8.1 'AGENCY PROBLEMS' THAT ARISE WHEN START-UPS SEEK EXTERNAL EQUITY INVESTORS – SEPARATING OWNERSHIP FROM CONTROL

Potential agency problems arise whenever an agent is given decision-making authority in a principal–agent relationship, but the agent's interests are not fully aligned with the principal's. In such situations, it should be expected that if both the agent and principal are utility maximizers, 'the agent will not always act in the best interests of the principal.'[1] In the case of external equity investments, the managers of the company (i.e., the agents) may have differing interests from the external shareholders (i.e., the principals).[2] For example, the managers may be incentivized to spend the company's resources on wasteful perquisites (items that do not generate direct financial returns to the company, such as lavish offices, first-class travel or private club member-ships, but do provide non-pecuniary benefits to the recipients of the perquisites) or fraudulently appropriate for their own private benefit a portion of the company's wealth. In either case, the manager will enjoy 100% of the fruits of such misbehavior, but bear only a small percentage of the cost since the managers own only a portion of the company's equity.[3] This dynamic may also encourage a manager to work at a less than optimal level (e.g., the manager may prefer to spend time golfing rather than building the business), because the manager receives the full benefit of her shirking while once again only bearing a fraction of the cost.

Notes:

1. Michael C. Jensen and William H. Meckling (1976), 'Theory of the Firm: Managerial Behavior, Agency Costs and Ownership Structure', 3 J. of Fin. Econ. 305, 308 (1976).
2. *Id.* at 309, 312–313.
3. *Id.* at 312–313; *see also* Paul Gompers and Josh Lerner (2006), *The Venture Capital Cycle* (2nd ed.), 159; Daniel Sandler (2004), *Venture Capital and Tax Incentives: A Comparative Study of Canada and the United States*, 14.

- *Professional venture capital firms*: these tend to be professionally-managed funds. Outside investors – frequently the more sophis-ticated traditional equity investors – commit money to a venture capital fund, which is managed by professionals ('venture capital managers') who are charged with 'finding, funding and assisting'[19] young start-ups. In addition to such diversely-funded venture capital funds, the classification of professional venture capital firms also includes dedicated venture capital firms that are primarily funded by a single corporation, a university or a government entity. In each case, however, there is a professional venture capital manager managing the funds – even if the funds come primarily from a single source.

- *Angel investors (or business angels):*[20]   refers to wealthy private individuals who invest their own capital directly in start-ups, rather than rely on a financial intermediary such as a professional venture capital manager.[21]
- *Private equity firms:*   private equity firms are frequently referred to as 'buyout' or 'leveraged buyout' firms. Their primary function is to acquire all, or a majority, of an underperforming, but asset-rich, operating company. The operating company can be publicly-traded or privately-held. The private equity firm will seek to resolve whatever problems are causing the operating company to underperform – e.g., replace management or restructure the company. Some of the more well-known private equity firms include Kohlberg Kravis Roberts (KKR), the Blackstone Group, the Carlyle Group, and Texas Pacific Group.

This chapter focuses on professional venture capital firms, rather than angel investors or private equity firms, because it is the professional venture capital managers who have demonstrated the greatest skill at resolving the information and management problems that plague start-ups.

1.2.1   Identifying worthy start-ups: the pre-investment screening process
The most obvious contribution that venture capital managers make to the start-up industry is their ability to overcome the pre-investment information problems – i.e., uncertainty about future success and severe information asymmetries. Before we continue, the term 'uncertainty' needs a little more explanation. The concept of 'uncertainty' is often confused with the term 'unknowable.' In fact, they are very different concepts.[22] When something is unknowable, 'no amount of research or analysis' will improve one's ability to divine the eventual result.[23] With regard to uncertainty, however, research, analysis and experience can help to improve one's ability to divine the eventual results and thereby reduce the uncertainty. In essence, uncertainty can be reduced and seemingly random outcomes can be intelligently dealt with. Much of successful start-up investing involves identifying and reducing the uncertainty surrounding start-ups.

To reduce pre-investment information problems, venture capital firms tend to conduct a very detailed screening process prior to agreeing to invest in a start-up. This screening process, typically referred to as 'due diligence,' involves an intensive review of the start-up's investment worthiness, including:

- The start-up's management team and its ability to successfully run the company;
- The start-up's business plan, with a particular focus on the size, and potential for growth, of the start-up's targeted market, the executability of the plan, potential competitive advantages (e.g., barriers to entry to the market) that the start-up may possess and the start-up's competitors;
- The quality of the product or service being offered by the start-up,[24] including through discussions with current customers;
- The strength of the start-up's intellectual property rights;[25] and
- A detailed review of the start-up's financial projections.[26]

This screening process allows the venture capital manager to gather the information necessary to make an informed investment decision and to eliminate those ventures that are unlikely to generate sufficient returns or whose management teams will require an inappropriate amount of monitoring.

1.2.2 The need for venture capital firms to be 'active' investors

Identifying worthy start-ups in which to invest is only a partial solution for investing successfully in such companies. Many of the severe information and agency problems that make start-up investments so challenging persist after an investment has been made. To adjust to the post-investment challenges posed by start-ups, venture capital firms tend to be active, rather than passive, investors. In addition to providing their portfolio companies with investment capital, venture capital firms also tend to provide their portfolio companies with a variety of valuable 'non-financial' services that help to reduce start-up information and agency problems and improve the performance of these companies.[27]

- *Management assistance:* venture capital managers typically possess substantial expertise in the industries in which they invest. They have experience moving similar companies 'up the development path from the startup stage' and possess specialized 'market knowledge based on other investments in the portfolio company's industry and related industries.'[28] This experience allows the venture capital managers to assist their portfolio companies with a range of management tasks from advising and mentoring (e.g., helping them through the predictable problems that start-ups face as they build their companies) to identifying and recruiting necessary management and other key personnel as the company grows.[29]

- *Monitoring and control*:   venture capital managers generally insist on receiving meaningful representation on the board of directors of their portfolio companies. This board representation provides the venture capital managers with the ability to influence/control the behavior of the company's managers and to reduce information asymmetries because most company law systems, including China's, provide that a company's officers answer to the board.[30]

- *Reputational capital*:   investors are not the only parties that suffer from severe informational problems when dealing with start-ups. Customers, strategic business partners, talented employees and suppliers also need to determine whether a particular start-up is sufficiently strong to warrant a commitment of their time and resources.[31] Venture capital firms can help to reduce this problem for their portfolio companies by serving as a 'reputational intermediary' on their behalf.[32] In effect, the venture capital firm 'certifies' or 'lends its strong reputation' to its portfolio companies by identifying them as (i) having survived the arduous venture capital screening process and (ii) being subject to continuous venture capital monitoring. Not surprisingly, start-ups with highly-respected venture capital firm investors are taken more seriously by customers and strategic business partners, are better able to attract talented employees, and receive more favorable treatment from suppliers.[33]

While many companies can benefit from such non-financial services, start-ups are in particular need of such services. For more established companies, such value-added services will typically be either less necessary – due to the greater maturity of the company (e.g., management assistance should be less valuable and the ability to project the company's future financial performance should be less uncertain) – or provided by an array of other sources (e.g., various financial intermediaries such as research analysts, investment banks, public auditors, or mutual fund managers[34]). For start-ups, venture capital firms may be the sole meaningful provider of such non-financial services.

2.   CHINA'S VENTURE CAPITAL SYSTEM

Over the last 25 years, China has laid the groundwork for a modern venture capital system. During that timeframe, venture capital has gone from being a minor, state-controlled financial experiment (which initially failed) to a significant source of start-up capital that embraces many of the principles of Western venture capital systems. While still developing, venture capital has become a $10+ billion a year business in China.[35]

2.1 Initial Efforts to Develop a Venture Capital System (1984–1998)

China's S&T efforts during the 1980s were successful at spurring the creation of new technology ventures, but not necessarily encouraging their growth and development. 'Creating' new start-ups is only a partial solution for a vibrant start-up community. Start-ups also need capital and non-financial services to allow them to grow and develop, and China did not have the infrastructure to provide its newly-created start-ups with such capital or non-financial services. There were no 'professional' sources of early-stage capital at that time. GRIs and research universities provided early-stage capital on occasion, as did banks and various central and local government sources, but each source tended to be constrained in the level of surplus funds it could commit to start-ups and none had the expertise that one expects from professional venture capital firms.[36]

In the mid-1980s, China began efforts to develop a venture capital industry that could serve the needs of China's emerging high-technology start-ups.[37] China's early efforts at venture capitalism, however, struggled.[38] During the mid-1980s, the focus of venture capital in China was infrastructure and property investments, such as hotels and tourist facilities.[39] These early investments were not overly successful, which reduced the appetite of many early venture capitalists.[40] The interest in venture capital investments began to return in the late-1980s, but the focus of such investments was still not on high-technology start-ups.

> The government still desired such investments and China's steady economic growth encouraged venture capital investment. However, the goals of the government and venture investors were in conflict. The central government wanted high tech and infrastructure investment, while at the time, investors preferred to focus on lower technology industries with lower risks . . . As a result, there were few venture capital investments in high tech startups.[41]

While specific and accurate statistics are difficult to obtain, available data support the notion that China's venture capital industry was not robust through the mid-1990s. In 1997, for example, statistics from Zero2IPO, a consulting firm based in Beijing, found that only seven venture capital investments were conducted in China.[42] The investment amount was disclosed in only four of those deals, and the total investment amount for those four deals was US$31.1 million (or US$7.78 million per

Table 8.2   Early efforts in China's venture capital industry[1]

Year	Event	Description
1984	State Science and Technology Commission (SSTC) study suggested that a Chinese venture capital industry be developed.	The SSTC's S&T Development Research Center conducted a study on how to develop high-technology industries in China. The study suggested that a venture capital system be developed to support China's high-technology industries.[2]
1985	Decision on Reform of the S&T Management System	The Decision was issued by the Central Commission of the Communist Party and the State Council. In addition to trying to create a 'technology market' (see Chapter 2),[3] the Decision proposed that venture capital be established to support risky, high-technology, rapid-growth start-ups.[4] This was the Chinese government's first official statement on venture capitalism.[5]
1986	China New Technology Venture Capital Company Ltd (CNTVCC) began operations.	CNTVCC was China's first venture capital firm,[6] and provided a variety of venture capital and incubator services.[7] SSTC and the Ministry of Finance (MOF) co-funded CNTVCC, with SSTC holding 40% of the shares and MOF holding 23%.[8] CNTCVV provided a variety of venture capital and incubator services.[9] CNTVCC has since closed.[10]
1988	Torch Program was established.	The Torch Program played a substantial role in China's initial venture capital efforts that is described in detail in this chapter.
1989	China Ke Zhao Tech Co., Ltd. (CKZTC) was established.	Four companies attached to the Ministry of Science, National S&T Industry Commission and Merchant Bureau funded China's second venture capital company, and its first Sino-foreign venture capital company.[11] CKZTC sought to promote the commercialization of S&T achievements that resulted from various government sponsored R&D programs like the Torch Program and Project 863.[12]
1991	Provisional Measures for China's High-Technology and New Technology Industry Development Zones	Issued by the State Council, the provisional measures permitted venture capital activities in high-technology development zones.[13]

Year	Event	Description
1992	Various venture capital firms were established in Shenyang, Chongquing, Taiyuan, Jiansu, Zhejian, Guangdong and Shanghai.	Following the '1991 Provisional Measures,' a number of development zones established venture capital firms.

Notes:
1. This table was motivated by similar tables produced by Jonsson. Yuiya Li (2005), *Investing in China: The Emerging Venture Capital Industry*, 33–52; and Steven White, JIan Guo and Wei Zhang (2004), 'Financial Systems, Investment in Innovation and Venture Capital: The Case of China', in Anthony Bartzokas and Sunil Mani (eds), *Financial Systems, Corporate Investment in Innovation and Venture Capital*, pp. 163–168.
2. *Id.*, at 163. LI, *supra* note 1, at 33. Jack C. Fensterstock and Aimin Li, (2000) *Status of Venture Capital in China* 2, available at www.mcbc.net/Articles/China%20VC%20%20 Status.pdf.
3. *See* Chapter 2, Section IV.
4. A copy of the Decision on Reform of the S&T Management System is available (in Chinese) at www.edu.cn/documents_8573/20090908/t20090908_405838.shtml.
5. Fensterstock and Li, *supra* note 2, at 2. Li, *supra* note 1, at 33.
6. White, Gao & Zhang, *supra* note 1, at 163 and 181.
7. Li, *supra* note 1, at 33.
8. *Id.*
9. *Id.*
10. *Id.*
11. *Id.*, at 33–34.
12. *Id.* The Pragma Corporation (2002), Final Report: Asian Development Bank TA 3534-PRC – Development of SME Financing Support System, Vol. 1, 131, available at www.adb.org/Documents/Reports/Dev_SME_Fin_System/sme_v1.pdf.
13. Article 6(3) of the Provisional Measures for China's High-Technology and New Technology Industry Development Zones, available at http://law.gaotai.gov.cn/ englishlaw/article/2005-11-10/904-1.htm.

deal).[43] In 1997, the early struggles of the Chinese venture capital industry were highlighted by the high profile bankruptcy of China New Technology Venture Capital Company Ltd (CNTVCC),[44] which was China's first venture capital company.[45]

2.1.1 Government-backed venture capital firms (GVCs)

There are a variety of explanations for China's early venture capital struggles. One of the more compelling explanations is that China's early venture capital firms were not truly 'venture capitalists' as one thinks of that term in the United States. China's early venture capital firms were controlled by the government (government-backed venture capital firms, or GVCs) and tended to operate more like government agencies than they did like disciplined, profit-seeking venture capitalists.[46] Consider, for example, CNTVCC. CNTVCC was established by the central government in 1985 and began operations in 1986,[47] and was China's first venture capital company. China's State Science and Technology Commission owned 40% of CNTVCC and the Ministry of Finance owned 23%.[48] Structurally, CNTVCC was a state-owned enterprise (SOE) that answered to the state, not a venture capital fund answering to individual investors. This government influence resulted in CNTVCC operating similarly to a 'central government agency with the mandate to support national technology venture policy objectives, rather than a profit-oriented private enterprise.'[49] While CNTVCC was a vehicle of the central government, later GVCs tended to be established and controlled by local governments, 'usually led by the local bureau of the science and technology commission and supported by the finance department of the local government.'[50]

What these GVCs had in common, whether central government GVCs or local government GVCs, was that either they were majority financed by government entities or the government served as the majority stockholder.[51] There were some benefits from this government-controlled structure. The largest benefit was the linkages that formed between the GVCs and the start-ups that were located in the various government-sponsored Development Zones (see Chapter 2, Section 3.3). The GVCs' government status provided them with 'preferential access to information and investment opportunities'[52] for Development Zone ventures. Overall, however, these benefits were far outweighed by a variety of problems that resulted from the government's early domination of the venture capital industry:

- *Investment decisions were compromised*: one of the biggest advantages of U.S.-style venture capital funds is their disciplined, profit-centric approach to investment decisions. GVCs were unable to fully imitate that disciplined approach. For GVCs, investment decisions were greatly influenced by government policy and the interests of local government leaders, rather than being driven purely by profit merits.[53] These outside influences have a tendency to encourage venture capital managers to pursue bad investments and to dissuade such managers from ruthlessly terminating failing projects.[54]
- *Regulatory environment failed to evolve*: venture capital requires a highly-evolved legal system to operate optimally. This topic will be treated more fully in Section 3 below, but for now we will simply point out that U.S. venture capital funds have developed a number of practices (Modern VC Practices) that help such funds to be more economically efficient, but require regulatory support to implement.

For example, U.S. venture capital funds tend to be structured as limited-duration limited partnerships; their investments tend to be in the form of preferred stock; and they encourage their portfolio companies to use equity compensation for their employees. The profit demands of sophisticated private investors who invest in the venture capital funds have helped to shape these Modern VC Practices, and U.S. legal infrastructure has co-evolved with these practices to legitimize them. Because early GVCs answered to government-backers, rather than sophisticated private investors, there was reduced pressure to develop Modern VC Practices and the more complex legal infrastructure that is needed to legitimize those practices.

- *Reduced supply of venture capital*: without the legal infrastructure to permit the creation of private venture capital funds, private investors (who make up the majority of venture capital fund investors in the United States) were largely excluded from putting their capital to work for start-ups. Privately-funded venture capital funds did not begin to significantly proliferate in China until 1998 following the 'No. 1 Proposal' (see Table 8.3).

- *Lack of talented venture capital managers*: one of the biggest challenges that any country faces when first trying to build a vibrant venture capital industry is to develop a pool of talented individuals that can serve as venture capital managers. Competent venture capital managers require a highly-specialized skill set that is not easy to develop. Venture capital managers need a multidisciplinary skill set that allow them to identify (and get access to) the most attractive start-ups and, once an investment is made, to both monitor and nurture the company. Typically, such skill set involves an in-depth understanding of finance, management, the technology that is core to the start-up, the industry in which the start-up operates, and the legal and regulatory issues (e.g., IP issues) that impact the start-up. Such a skill set takes years to develop and the complexity of the skill set leads to a great diversity in competence among venture capital managers. Competent managers, even in established venture capital industries like the United States', tend to be in short supply.[55] China's early GVCs faced a number of problems with their venture capital managers. First, there was no ready pool of talented managers to draw from. Second, the government structure of the GVCs reduced the financial remuneration that was available to managers – as compared to private venture capital funds – which likely made the position less attractive to the most talented individuals. Finally, the government structure of the GVCs did not ruthlessly eliminate weaker managers as effectively as the U.S.-style system does. In

Table 8.3   Further development/understanding of venture capital in China[1]

Year	Event	Description
1995	Decision on Accelerating S&T Progress	The Central Committee of the CPC and the State Council issued the decision which, among other things, emphasized the need to develop a venture capital system to support China's S&T R&D efforts.
1996	Law Promoting the Transformation of S&T Achievements	NPC passed this law, which was the first 'law' to encourage the establishment of venture funds and allow venture capitalism as a commercial activity.[2]
1996	SSTC delegation sent to United States	SSTC sent a delegation of scholars to the United States to study the U.S. approach to financing smaller companies, IP rights and venture capitalism.
1997	Deng Nan, Vice Minister of SSTC and daughter of Deng Xiaoping, was appointed to oversee major study of China's venture capital system.	Deng Nan worked in collaboration with a group from Tsinghua University's School of Economics and Management to provide a report with 'practical recommendations' for establishing a venture capital system in China.
1998	Fourth Meeting of National S&T Leaders Team.	Premier Li Peng chaired this meeting of China's leading S&T policy group. The group concluded that a blueprint for a Chinese venture capital system should be developed and put into action.
1998	Proposal for Developing Venture Capital Industry in China (commonly referred to as the 'No. 1 Proposal').	This proposal was submitted by the China National Democratic Constructive Association at the Ninth National Committee of the Chinese People's Political Consultative Conference. It was listed as the 'No. 1 Proposal' for the conference, which is where it got its popular nickname. The No. 1 Proposal generated a lot of interest in venture capitalism from China's top leaders and led to the creation of a substantial number of new 'corporate-backed' venture capital firms. The No. 1 Proposal was a true watershed event for China's venture capital industry.
2006	Provisional Measures for the Administration of Venture Capital	The VC Measures are the first national statute to govern non-foreign venture capital investing in China. Foreign-invested venture capital enterprises (discussed below in section 2.2.3)

Year	Event	Description
	Enterprises (the VC Measures).	continue to be governed by separate regulations that were adopted in 2003 (see Table 8.5). The VC Measures govern the formation, business operations of, and policy support for venture capital firms operating in China.

Notes:
1. This table was motivated by similar tables produced by Jonsson Yinya Li (2005), *Investing in China: The Emerging Venture Capital Industry*, 33–52 and Steven White, Jian Gao and Wei Zhang (2004), 'Financial Systems, Investment in Innovation and Venture Capital: The Case of China', in Anthony Bartzokas and Sunil Mani (eds), *Financial Systems, Corporate Investment in Innovation and Venture Capital* 163–168. The tables produced by Li and White, Gao and Zhang also served as the primary sources.
2. Article 24 of the Law Promoting the Transformation of S&T Achievements.

the United States, most venture capital firms are structured as self-liquidating funds,[56] which means the venture capital firm's investment capital is invested only once, with any returns distributed immediately to the venture capital firm's investors rather than being reinvested by the venture capital firm.[57] Fund managers who wish to continue in the venture capital business (which appears to describe most fund managers) must raise additional capital and create a further fund, and will typically need to show a strong track record of success in order to raise any future funds.

2.1.2   Bank-centric approach to financing start-ups

The generally weak, GVC-dominated venture capital industry in China during the 1980s and through much of the 1990s resulted in a far from efficient dynamic for Chinese start-ups. GRIs and universities would provide both the original technology and seed capital for many newly-formed start-ups.[58] The new start-ups would then be located in China's Development Zones, which provided fundamental incubator services such as physical space and basic infrastructure.[59] One estimate suggests that GRI-backed start-ups accounted for roughly half the ventures operating in China's Development Zones at the outset of the 1990s.[60]

The greatest value of the Development Zones was the valuable certification function they provided for the start-ups. First, the Development Zones helped the start-ups to qualify for preferential treatment under the Torch Program and certain other government policies, which signaled to banks which start-ups should receive loans.[61] Once the Torch Program was up and running, banks quickly became the primary source of funding for start-ups. In 1988, when the Torch Program was initiated, banks accounted for roughly 10% of financing to start-ups.[62] In 1990 and 1991, however, banks accounted for 50% and 70% of start-up financing respectively.[63] Banks appear to have been responding to the government's policy to support start-ups, and the Torch Program was its signal for identifying which start-ups should be supported.[64] Bank financing, however, was largely limited to later-stage expansion financing for start-ups, and tended to be absent from the seed capital and early-stages of start-up investment.[65] The dominant role of banks in start-up financing appears to have continued through the mid-1990s in China.[66] The Development Zones also signaled to the GVCs and local government officials – who influenced the GVCs – which start-ups should receive the limited venture capital funds.

China's bank-centric approach to financing start-ups proved to be highly problematic. First, the supply of seed and early-stage financing for start-ups was too small, as it was largely limited to the resources that GRIs and universities could dedicate to the task.[67] Second, the weak screening process used by banks to provide loans to inherently high-risk ventures could not be a long-term strategy for China's banks. In the United States, start-ups wishing to receive expansion capital must typically go through a comprehensive screening process by one, or more, venture capital funds (see earlier discussion in section 1.2.1). The screening process is very arduous and few businesses that submit funding requests to U.S. venture capital funds actually receive funding.[68] In China, during the late 1980s and much of the 1990s, this methodical venture capital screening process was replaced by the much weaker screening process required for gaining access to a Development Zone and getting designated as a Torch Program company.

2.2 Developing a Modern Venture Capital Industry in China (1998–Present)

Fortunately for its start-up industry, China's understanding of venture capital and its optimal role in the innovation process evolved substantially during the 1990s. In 1998, the No. 1 Proposal that was made at the Ninth National Committee of the Chinese People's Political Consultative Conference (see Table 8.3) signaled a major shift in the government's understanding of venture capitalism and how it should be regulated, and marked the commencement of a modern venture capital industry in China. At the broadest level, the No. 1 Proposal sought to develop a Chinese venture capital industry based on Western models of private venture

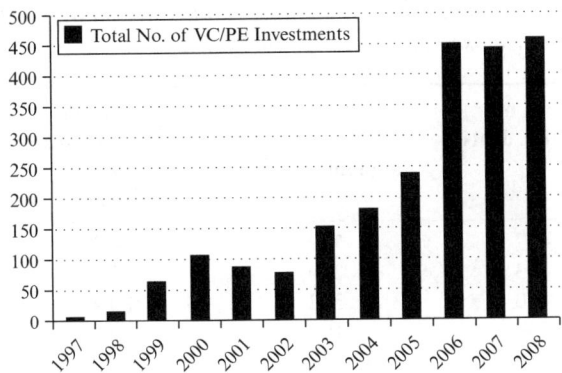

Source: Zero2IPO database.

Figure 8.1 *VC/PE investments in China: annual no. of deals from 1997–2008*

capital firms, rather than treating venture capital as just another form of government subsidization for new technology firms.[69] New forms of venture capital firms were created – namely corporate venture capital firms – and 'a wave of foundings involving government, corporate and foreign [venture] capital'[70] began in 1999. China now has four major categories of professional venture capital firms: (1) corporate-backed venture capital firms; (2) university-backed venture capital firms; (3) foreign venture capital firms; and (4) GVCs.

Figure 8.1 and Table 8.4 provide some insight into the dramatic growth of China's venture capital industry beginning in 1999. Before presenting and analyzing the data, however, we need to mention two important caveats. First, collecting data on a country's venture capital industry is an inherently challenging endeavor. Unless a venture capital firm is a listed company, it is not required to publicly disclose information about its investments, which severely hampers the accuracy of any data collection efforts. As a result, the data presented in Figure 8.1 and Table 8.4 should be treated as general estimates of venture capital activity in China, rather than hard data. Second, data sources for venture capital activity in China tend not to make a distinction between 'venture capital' and 'private equity' (e.g., leveraged buy-out firms) when collecting and presenting their data. The data in Figure 8.1 and Table 8.4 therefore include both venture capital and private equity activity. Because there tends to be little leveraged buy-out activity in China,[71] the data remain helpful for understanding the level of venture capital activity in China.

Table 8.4 *VC/PE investments in China: annual amount invested (based on disclosed amounts) from 1997–2008*

	No. of VC/PE deals	No. of deals that disclosed amount invested	% of deals with disclosed amount invested	Disclosed amount invested (in millions)
2008	457	339	74.2	$11 177.2
2007	443	320	72.2	$11 645.4
2006	450	362	80.4	$13 138.8
2005	239	193	80.8	$2523.4
2004	180	142	78.9	$1441.9
2003	152	123	80.9	$1409.8
2002	78	49	62.8	$445.9
2001	87	46	52.9	$1637.8
2000	106	59	55.7	$416.6
1999	65	47	72.3	$358.8
1998	16	5	31.3	$18.3
1997	7	4	57.1	$31.1

Source: Zero2IPO Database.

2.2.1 Corporate-backed venture capital firms (CVCs)

CVCs quickly began to emerge in China following the No. 1 Proposal. The first two CVCs – Beijing High-Tech Venture Capital Ltd. and Beijing Sci-Tech Venture Capital Company Ltd. – were founded in October 1998.[72] These two initial CVCs benefited from strong government backing, which caused a perception that they were still largely GVCs.[73] By 1999, truly corporate-backed CVCs were being formed and CVCs soon constituted the majority of non-foreign venture capital firms in China.[74] CVCs tend to have a variety of investors, including government entities, but their primary source of funds tends to be publicly-traded Chinese companies.[75]

Looking ahead, a significant issue to monitor for CVCs will be the professional backgrounds of their venture capital managers. While this is a bit of an over-simplification, venture capital managers tend to come from either an 'operational' or a 'financial' background. Managers with an operational background tend to be former entrepreneurs with experience building and exiting successful start-ups or leading industry executives. Managers with a financial background tend to be former investment bankers, or they have some other type of investment experience, and have little to no experience operating a company. We would be concerned if CVCs over-relied on managers with financial experience.

An over-emphasis on finance-oriented managers can negatively impact the ability of the venture capital firms to develop and nurture start-ups. A lack of industry/operational expertise makes it more difficult to accurately identify which start-ups have the greatest potential to succeed and to provide practical mentoring for companies in the venture capital firm's investment portfolio. While this is a very broad generalization, there is a belief in venture capital circles that venture capital managers with industry/operational expertise are more likely to take a longer-term, value-added development role with their portfolio companies, while those with a predominantly finance background tend to take a shorter-term view. Rather than build companies for the long term, the greater emphasis for finance-oriented managers is on seeking quick, profitable exits from their investments. Unfortunately, the pool of former investment bankers tends to be much deeper than the pool of successful former entrepreneurs, so the risk of over-reliance on finance-oriented managers is a very real one.

2.2.2 University-backed venture capital firms (UVCs)

In 2000, universities meaningfully began to enter the venture capital market by creating their own university-backed venture capital firms.[76] As with CVCs, UVCs are likely to have a variety of investors, but the host university tends to play a substantial role in both funding and controlling the UVC. For the most part, UVCs limit their investments to opportunities that arise from their host universities.[77] On the positive side, the close relationship with the host university provides a UVC with privileged access to deal flow and a greater ability to obtain accurate information about the investment quality of their opportunities.[78] On the negative side, the ability for non-financial factors to creep into the UVCs' decision making process is substantial. The ability to ruthlessly terminate a weak project that is being led by a popular professor, for example, can be more difficult because of the closer relationship between the UVC and its portfolio companies.[79] Some researchers have also suggested that UVCs are less attractive career opportunities for the best venture capital managers, which reduces the managerial expertise at UVCs.[80]

2.2.3 Foreign venture capital firms (FVCs)

FVCs have been allowed to operate in China since the 1980s, but their early investment opportunities were severely limited by both regulation[81] and the limited supply of attractive investment projects.[82] Even though early FVCs had to contend with a complex and burdensome regulatory structure that was not overly conducive to FVCs, they still managed to enter China's venture capital market and eventually take a dominant role. By 2001, eight of the top 10 venture capital firms in China, and 14 of the

Table 8.5   Opening of Chinese market to China-domiciled FVCs

Year	Event	Description
2001	Interim Regulations on the Administration of Foreign-invested Venture Capital Enterprises	The interim regulations were jointly issued by the Ministry of Foreign Trade and Economic Cooperation (now the Ministry of Commerce), the Ministry of Science and Technology and the State Administration for Industry and Commerce. The interim regulation eliminated many of the prior regulatory barriers that hindered China-domiciled FVCs and allowed them to operate on a much more even level with domestic venture capital firms.*
2003	Regulations on the Administration of Foreign-invested Venture Capital Enterprises	Jointly issued by the Ministry of Foreign Trade and Economic Cooperation (now the Ministry of Commerce), the Ministry of Science and Technology, the State Administration for Industry and Commerce, the State Administration of Taxation and the State Administration of Foreign Exchange. These new 2003 rules replaced the prior interim regulations. A detailed discussion of the new 2003 regulations is set forth in this section 2.2.3.

Source: * Jonsson Yinya Li (2005), *Investing in China: The Emerging Venture Capital Industry*, 26.

top 20, were FVCs.[83] The new generation of FVCs tended to be stronger and have greater experience than their domestic counterparts. A 2002 study found that domestic venture capital managers averaged 2.1 years of relevant experience, compared to 11.9 years for FVC managers.[84] FVCs also tended to be more skilled at providing the valuable non-financial services that are such an essential part of venture capital investing in the United States.[85]

Presumably, the Chinese government recognized the greater value being played by FVCs because, beginning in 2001, it substantially improved the regulatory environment for FVCs. In 2001 and 2003, the Chinese government implemented regulations that allowed for FVCs to operate in China through China-domiciled venture capital enterprises (China-domiciled FVCs – also commonly referred to in the industry as RMB funds).

The 2003 Regulations on the Administration of Foreign-invested Venture Capital Enterprises (the 2003 FVC Regulations) went into effect on March 1, 2003 and immediately replaced the 2001 Interim Regulations on the Administration of Foreign-invested Venture Capital Enterprises. Some of the more important aspects of the 2003 FVC Regulations include:

- *Limited liability for FVC investors*: China-domiciled FVCs can be either incorporated or unincorporated entities.[86] For incorporated China-domiciled FVCs, the investors will have limited liability (they are only liable for the amount of their equity investment in the China-domiciled FVC).[87] For unincorporated China-domiciled FVCs, the investors are jointly and severally liable for the China-domiciled FVCs obligations, unless the China-domiciled FVC's 'mandatory investor' (e.g., its general partner) agrees to assume unlimited liability – in which case the other investors are only liable for the amount of their equity investment in the China-domiciled FVC.[88] Most of the China-domiciled FVCs established under the FVC Regulations have been unincorporated entities.[89]

- *Scope of business*: China-domiciled FVCs are permitted to make equity investments 'principally' in unlisted, high-technology enterprises.[90] Such equity investments may be used to start new companies, to purchase newly-issued stock of existing (but unlisted) companies or to acquire the equity interests of existing stockholders.[91] In addition to their investment activities, China-domiciled FVCs may also provide advisory and managerial services to their portfolio companies in order to generate capital gains from their equity investments,[92] and venture investment consulting services (including to other venture capital firms).[93]

- *Prohibited activities*: China-domiciled FVCs may not invest in publicly-traded companies, although they are permitted to maintain shares in portfolio companies that are subsequently listed.[94] China-domiciled FVCs are subject to a list of additional prohibitions. For example, they may not invest directly or indirectly in real estate, use leverage to make investments, invest money that is not owned by the China-domiciled FVC (i.e., China-domiciled FVCs may not serve as a conduit for investment capital that has not been committed to the China-domiciled FVC), or make most forms of loans or guarantees.[95]

- *Regulatory approvals*: the Ministry of Commerce must approve the establishment of all China-domiciled FVCs.[96]

- *Investors in China-domiciled FVCs and minimum capitalization*: China-domiciled FVCs must have between two and 50 equity investors, with at least one investor serving, and qualifying, as a 'mandatory investor' (described below).[97] The minimum aggregate capitalization for a China-domiciled FVC is US$5 million if it is incorporated, but US$10 million if it is unincorporated.[98] Capital contributions may be made in installments, although the period of capital calls is not to exceed five years.[99] Investors in the China-domiciled FVC may not reduce the amount of their subscribed capital contribution.[100]

- *Mandatory investors*:   mandatory investors serve a similar function to the general partners in the typical U.S. venture capital fund structure. To qualify as a 'mandatory investor', the investor must satisfy the following conditions: (1) venture capital investing is its primary business; (2) for the three years prior to the establishment of the China-domiciled FVC, it must have had at least US$100 million under management, of which at least US$50 million is dedicated to venture capital investments; and (3) it has at least three professional managers, each of whom has at least three years of experience in venture capital investments.[101] For unincorporated China-domiciled FVCs, the mandatory investor's capital contribution must be at least 1% of the subscribed capital and 1% of the actually committed capital. For corporate China-domiciled FVCs, the minimum threshold is 30% for subscribed and actually committed capital.[102]

- *Management of the China-domiciled FVC*:   China-domiciled FVCs must establish a joint management committee (for unincorporated FVCs) or a board of directors (for incorporated FVCs).[103] The committee/board may delegate management of the China-domiciled FVC's investments to an internal management structure or outsource it to another venture capital company.[104] The venture capital management firm may be a Chinese venture capital company, another China-domiciled FVC or an overseas venture capital company.[105]

- *Limited duration entities*:   China-domiciled FVCs are limited duration entities and must stipulate their term of existence, which is not to exceed twelve years, in their articles of association.[106]

2.2.4   GVCs

It is worth noting that GVCs have had to react to the modernization of China's venture capital industry. Local governments are no longer injecting fresh capital into GVCs, which has caused them to diversify their funding sources and/or generate returns on their investments in order to fund further investments.[107] For those GVCs that survive as viable venture capital entities, presumably it will be because they have transformed themselves into professionally-managed, profit-driven venture capital firms that are able to successfully compete with the CVCs, UVCs and FVCs for both investors and investment opportunities.

2.3 Need for Greater Experience/Expertise of Venture Capital Managers and Other Key Actors

Despite considerable progress, China's venture capital system – in particular the domestic portion of the industry – remains relatively immature when compared to Western systems[108] and has room for improvement. For example, there is a general perception that venture capital managers for China's domestic venture capital firms have less experience and expertise than their FVC counterparts.[109] A recent study by professors Tan, Zhang and Xia (2008) provides some empirical support to that general sentiment.[110] Tan, Zhang and Xia conducted extensive interviews of 35 major venture capital firms that were operating in China from 2000 to 2002 (22 were Chinese venture capitals firms and 13 were FVCs) to assess their use of control and incentive mechanisms. Among the study's primary findings were:

- *Less active monitors*: Chinese venture capital firms monitor their portfolio companies less actively than FVCs. While all of the study respondents required board seats (unless they were only acquiring a minimal stake), Chinese venture capital firms tended to hold fewer board meetings.[111] Chinese venture capital firms also required less frequent financial disclosure.[112] Table 8.6 provides select data from the Tan, Zhang and Xia study.
- *Weaker influence over management decisions*: Chinese venture capital firms also exert weaker influence over the managerial decisions of their portfolio companies than FVCs. For example, Chinese venture capital firms are less likely to 'stage'[113] their investments, use convertible preferred stock in their investments, or possess veto rights over major decisions. Table 8.7 provides select data from the Tan, Zhang and Xia study.
- *Provide fewer 'non-financial' services*: finally, Chinese venture capital firms tend to provide fewer of the valuable non-financial services (e.g., managerial assistance, monitoring and control, and reputational capital) that have helped venture capital firms, generally, to earn their reputation as 'value-added' investors. Based on their interviews, Tan, Zhang and Xia concluded that:

> [D]omestic venture capitalists provide much less to entrepreneurs in terms of value-added services. . . Indeed, an underlying difference between these types of firms is that domestic venture capitalists do not see addressing operational issues as an important part of their role as investors. Instead, they concentrate their monitoring and participation on the financial aspects of the investee firms.[114]

Table 8.6   Monitoring tendencies: Chinese v. foreign VC firms (select data from Tan, Zhang and Xia study (2008))

	Frequency of board meetings with portfolio companies					Frequency of financial disclosure from portfolio companies		
	Monthly	Every 2 months	Quarterly	Every 6 months	Annually	Monthly	Quarterly	Less than quarterly
Chinese VC Firms								
No. of firms	0	3	7	10	2	16	6	0
% of firms	–	14%	32%	45%	9%	73%	27%	–
FVCs								
No. of firms	2	3	6	2	0	12	1	0
% of firms	15%	23%	46%	15%	–	92%	8%	–

Source: Justin Tan, Wei Zhang & Jun Xia (2008), 'Managing Risk in a Transitional Environment: An Exploratory Study of Control and Incentive Mechanisms of Venture Capital Firms in China', 46 *Journal of Small Business Management* 263, 274–275.

While the Tan, Zhang and Xia study provides valuable insights into the behaviors and tendencies of venture capital firms operating in China, caution should be taken when interpreting the study's results. To begin with, the study suffers from many of the typical problems that plague questionnaires/surveys, such as small sample size. How much should we read into the fact that a few more FVCs prefer monthly over quarterly board meetings? In addition, the study was conducted in 2002, which pre-dates the major increase in Chinese venture capital activity that began in 2003 (see Figure 8.1). Tan, Zhang and Xia explain that one reason for the different behavior/tendencies between domestic and foreign venture capital firms is that the venture capital managers for domestic firms tend to have much less experience. At some point, they should close the gap on this experience disadvantage. We do not have an informed opinion on whether they have already succeeded in closing the experience gap, but we do caution against simply projecting the experience gap forward. In fact, there is a growing belief that domestic funds have become very competitive with FVCs in what they can offer their portfolio companies.[115]

Domestic venture capital managers are not the only actors developing valuable operating experience in the Chinese market. While rapidly

Table 8.7 *Ability to control portfolio companies: Chinese v. foreign VC firms (select data from Tan, Zhang and Xia study (2008))*

	Use of 'staged' investment strategy				Use of convertible preferred stock in investments				Hold veto rights over major decisions			
	Widely used	Partially used	Rarely used	Never	Widely used	Partially used	Rarely used	Never	Widely used	Partially used	Rarely used	Never
Chinese VC firms												
No. of firms	1	3	8	10	1	3	3	15	1	6	4	11
% of firms	5%	14%	36%	45%	5%	14%	14%	68%	5%	27%	18%	50%
FVCs												
No. of firms	2	5	3	3	10	3	0	0	9	3	1	0
% of firms	15%	38%	23%	23%	77%	23%	–	–	69%	23%	8%	–

Source: Justin Tan, Wei Zhang and Jun Xia (2008), 'Managing Risk in a Transitional Environment: An Exploratory Study of Control and Incentive Mechanisms of Venture Capital Firms in China', 46 *Journal of Small Business Management* 263, at 270, 273 and 276.

developing, China's venture capital market remains a relatively recent phenomenon, which means that most actors in the system could benefit from greater experience. FVC venture capital managers that may have loads of experience in the U.S. market may need time to learn and adjust to the Chinese market. Lawyers capable of serving entrepreneurs and venture capital firms in their transactions also need experience. Experienced lawyers are better able to structure and document optimal venture capital financing arrangements, and promote best practices and reasonable financing terms throughout the market as a whole.

Fortunately, each of these experience problems – domestic venture capital managers, FVC venture capital managers, and venture capital lawyers – is the type of problem that should naturally work itself out over time. In fact, with the amount of venture capital activity taking place in China, this problem should work itself sooner rather than later.

3.   ABILITY OF CHINA'S LEGAL SYSTEM TO SUPPORT VENTURE CAPITALISM

As we noted earlier, venture capital requires a highly-evolved legal system to operate optimally. In addition to the secure property rights and consistently enforceable contracts that are required for any market based activity, venture capital requires a legal system that performs an array of more particularized functions. Venture capital, by its very nature, involves the creation of value-creative, but frequently complex, arrangements between investors, money managers, entrepreneurs, and high-growth companies. For venture capital to flourish, the law must effectively support these various arrangements. This section will look at some of the more crucial functions that a legal system must provide for a venture capital market to succeed and examines the ability of China's legal system to perform those functions.

3.1   Allowing Venture Capital Firms to Operate

At the most fundamental level, venture capital investing needs to be a legally-permissible activity for a market to develop. While that observation may seem so obvious as to be irrelevant, it has been a relevant question in China until recently. Venture capital investing was first made lawful in China only 25 years ago, with the Decision on Reform of the S&T Management System (1984) (see Table 8.2), and that was on a very limited basis. China did not officially authorize the broader venture capital activities that most tend to associate with modern venture capitalism until 1998 with its No. 1 Proposal (see Table 8.3). Regarding FVCs, which are the most active venture capital investors in China, their ability to operate in China through China-domiciled FVCs was severely limited until 2001 (see Table 8.5). At this point, however, China's laws and regulations clearly permit venture capital investing by both domestic and foreign venture capital firms.

A related issue is the regulatory burden placed on venture capital firms wishing to operate in China. There are a variety of valid reasons for regulating venture capital firms. For example, regulating venture capital firms can help to:

- Protect unsophisticated investors from incurring an inappropriate level of risk in their investments;
- Increase the efficiency of the venture capital market;
- Reduce the transaction costs involved with venture capital investing by standardizing certain procedures (e.g., standardizing disclosure requirements); and

- Reduce the systemic risk to China's overall venture capital market of fraudulent venture capital firms that could deter start-ups from accepting venture capital or investors from investing in venture capital.

These benefits from regulating venture capital firms, however, do not come without a cost. Regulating activities generates compliance costs. In an ideal setting, regulations are 'cost effective,' meaning the benefits from the regulations exceed the compliance costs. Unfortunately, developing cost-effective regulations is an inherently difficult task. In the case of venture capital firms, if the compliance costs become too high, such firms may decide to reduce the amount of investments they make or eschew investing in China altogether in order to avoid the excessive compliance costs.

It will be important for Chinese regulators to monitor this 'cost effectiveness' issue carefully in the future. China's more onerous regulations for FVCs, for example (see section 2.2.3), run the risk of being cost ineffective and could be discouraging otherwise willing FVCs from investing in Chinese start-ups. There appear to be a few competing interests behind China's regulatory approach to FVCs. On the one hand, China wants to encourage the inflow of valuable venture capital financial and non-financial resources into the country to benefit its emerging start-up industry. On the other hand, there are political concerns that too many of the best investments could go to foreigners (e.g., selling state assets too cheaply to foreigners)[116] as well as economic concerns that a large influx of foreign capital could inappropriately inflate the valuation of Chinese start-ups as too many dollars chase too few deals. While this could appear to require policymakers to strike a rather delicate balance between these interests – and that is probably a common viewpoint among policymakers – we view the situation a bit differently.We believe the policymaking balance for FVCs should tilt very heavily towards encouraging the inflow of foreign venture capital investments, and that it really is not even a very close call on that issue.

One trap that policymakers can easily fall into when developing regulations for venture capital firms is to fail to adequately differentiate venture capital firms from other financial actors. For example, venture capital firms are frequently lumped in with hedge funds or leveraged-buyout firms, which are very different financial actors. Venture capital firms are not buyout firms. They are not looking to acquire majority stakes in undervalued asset-rich companies – which is an activity that warrants substantial government attention in countries with large numbers of state-owned enterprises that are for sale. Venture capital firms are looking to invest relatively modest sums of money – e.g., $2 million to $50 million – to acquire minority interests in start-ups. Venture capital firms are not looking to turn round (or raid the assets of) a former state-owned enterprise. They are looking to acquire a minority interest in a rapidly-growing start-up, and then exit that investment in a reasonable time-frame (typically two to five years). In addition, the venture capital industry does not have the type of scale that should cause policymakers to fear an inflationary impact from their activities. Venture capital firms control a tiny fraction of the funds that are controlled by hedge fund and buyout industries.

One factor that appears to help FVCs as they manage the cost-effectiveness of China's regulatory environment is the cooperative treatment they receive from local governments. FVCs have reported that local governments, which are almost always in need of the type of risk capital and management expertise provided by venture capital firms to grow their economies, tend to be highly supportive of the venture capital industry in general and do not appear to have any bias against FVCs.[117]

3.2 Sources of Capital

As a general rule, venture capital firms are professionally-managed funds that invest other people's money. Outside investors commit money to the venture capital firm, and the firm's venture capital managers take that money and invest it in promising start-ups. The ability of venture capital firms to finance and support start-ups is therefore entirely contingent on their ability to find these outside investors that are willing to commit money.

The regulatory environment can play a big role in venture capital firms' ability to obtain investment capital. In the United States, for example, a change in pension fund regulations was one of the biggest reasons for the growth of venture capital since the early 1980s. The Employee Retirement Income Security Act of 1974 (ERISA) is a federal law that establishes a set of standards for most U.S. private pension plans.[118] One of the standards that ERISA establishes is to provide 'fiduciary responsibilities' for persons who manage and control a plan's assets, one of which is to manage the plan's assets with the care of a 'prudent man' (i.e., carefully and conservatively).[119] During the 1970s, most plan managers were concerned that investing in venture capital firms was too risky and would violate the 'prudent man' standard. In 1979, the U.S. Department of Labor ruled that portfolio managers could consider portfolio diversification in determining the prudence of a particular investment.[120] The implication was that pension funds could allocate a portion of their portfolios to higher risk

Table 8.8   *Comparison of Chinese domestic sources of venture capital with U.S. sources of venture capital*

Capital source	Chinese domestic sources of venture capital (2005)	U.S. sources of venture capital (2002)
Government	33%	0%
Industry	48%	10%
Financial Institutions	13%	10%
Pension Funds	0%	40%
Mutual Funds	0%	5%
Individuals	3%	21%
Other	3%	14%

Source: Lu Haitian, Jan Yi and Chen Gongmeng (2007), 'Venture Capital and the Law in China', 37 *Hong Kong Law Journal* 229, 244 at 244 (citing China Venture Capital Research Institution for the Chinese statistics and Thompson's Venture Economic Committee for the U.S. statistics).

investments, such as venture capital firms. Following the ruling, pension funds, which are one of the largest sources of investment capital in the United States, soon became the largest source of capital for U.S. venture capital firms.

In China, many of the traditional funding sources for venture capital firms have historically been prohibited from making venture capital investments. Chinese commercial banks,[121] insurance companies,[122] and securities companies,[123] and the National Social Securities Fund[124] have all been barred from making venture capital investments. These major capital sources have been highly restricted in the investment choices available to them – which included their ability to invest in venture capital. In addition to limiting the supply of venture capital, these restrictions have also led to a less than ideal mix of capital sources for the Chinese venture capital industry (see Table 8.8).

The problem with China's mix is the relatively low percentage of 'independent' sources of capital. In this case, an independent source of capital is one that is driven almost entirely by profit motives. An independent source of capital provides its capital to venture capital managers with the sole instruction being to find and nurture successful start-ups that will generate profitable investment. Venture capital that comes from financial institutions, pension funds, mutual funds, and individuals will typically be 'independent.' Independent sources of capital are preferable for a venture capital industry, since such sources allow venture capital managers to perform their function free from conflicts or demands that may come from 'conflicted' capital sources. Venture capital that comes from government or industry sources – which have historically been the primary sources of venture capital in China – are classic examples of conflicted capital. Government and industry sources of venture capital are more likely to have conflicting interests (e.g., job creation or local economic development (for government sources); or assisting potential customers (for industry sources)) that can negatively influence the venture capital managers' decisions and performance. You may recall the story of Boston University and Seragen (see Chapter 6, section 2.3.2) as an example of the damage that such conflicts can cause.

Over the last few years, there has been some loosening of the overall investment choices available to China's major capital sources, including their ability to invest in venture capital firms (see Table 8.9). It is too early to tell what the impact of this loosening will be. Will it help to propel China's domestic venture capital industry in a way that is similar to what occurred in the United States following the loosening of its ERISA regulations by allowing large amounts of 'independent' capital to flow into the venture capital market? Or will it cause too much venture capital to flood into China's venture capital market without sufficient infrastructure to handle the funds? We do not have the answer to these questions at the moment, but they are issues that should be monitored.

3.3   Onshore v. Offshore Investing Structures

One of the most important trends in Chinese venture capitalism over the last decade has been the prevalence of offshore investing structures. FVCs have developed as a dominant source of venture capital funding in China, and most of those FVC deals have been conducted as offshore transactions.[125] While the offshore strategy has been helpful for the development of China's venture capital industry, certain abusive practices came to the attention of China's government which has resulted in stricter regulations of the strategy and helps to explain the more recent increase in 'onshore' investing strategies.[126]

3.3.1   The basic structure for an offshore FVC transaction – the 'red-chip' strategy

The offshore structure can take many forms, but the simplest structure involves creating an offshore holding company that owns 100% of the Chinese start-up (see Figure 8.2). As the sole owner of the start-up, it is relatively easy for the holding company to structure the arrangement to ensure that it has complete control over the Chinese start-up. FVCs can then invest in the holding company, rather than directly in the Chinese start-up, with the FVCs' control over the holding company providing

Table 8.9   Loosening of restrictions on major capital sources investing in venture capital firms

Year	Event	Description
2006	Council of China Social Securities Fund invested RMB 1 billion in Bohai Industry Investment Fund.[1]	Bohai Industry Investment Fund invests in unlisted, high-growth start-ups.[2]
2006	S&T Guidelines (2006–2020) and resulting policies (see Chapter 7)	• Circular on Medium- and Long-Term Planning for S&T Guidelines (effective Feb. 14, 2006), which allowed securities firms to develop venture capital business. • Statement that insurance companies would be allowed to develop venture capital business.[3]
2006	China Banking Regulatory Commission issued: • Detailed Rules for the Implementation of the Policies on Policy Finance for Supporting Major National Scientific and Technological Projects; and • Guiding Opinions for Commercial Banks to Improve and Strengthen Financial Services to Hi-Tech Enterprises	These policy documents provided banks greater freedom in their ability to serve the venture capital industry (e.g., domestic venture capital firms can leverage from banks) and could be a signal that banks will eventually become sources for venture capital.[4]

Notes:
1. Lu Haitian, Tan Yi and Chen Gongmeng (2007), 'Venture Capital and the Law in China', 37 *Hong Kong L. Jour.* 229, 248–249.
2. *Id.*, at 249.
3. *Id.* (citing an interview from Cao Wendong, Associate Head of the Financing Department, National Development and Reform Commission, at the Strait Youngs Forum on June 16, 2006).
4. *See id.*

them effective control over the Chinese start-up as well. Because the holding company is really just a 'shell', the funds invested in the holding company are then transferred to the Chinese start-up through a capital contribution by the holding company or a loan.

Such offshore structures provide FVCs with a number of significant regulatory advantages when investing in Chinese companies:

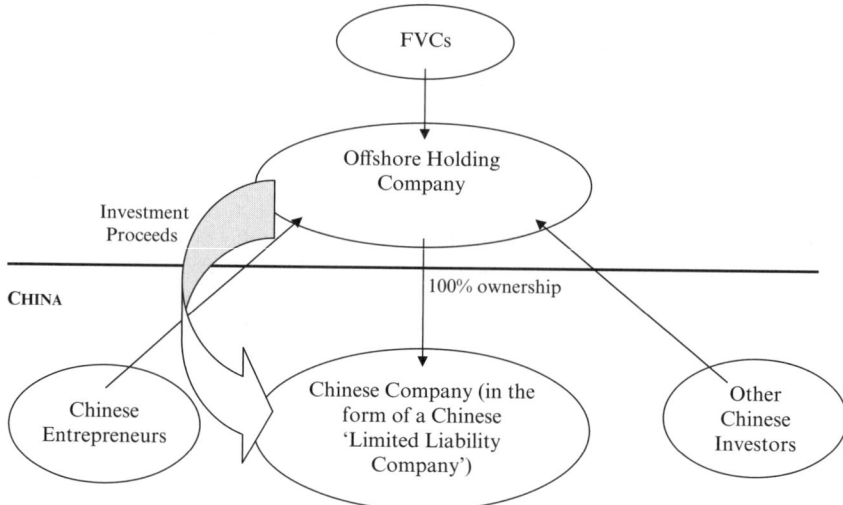

Figure 8.2   Typical offshore investing structure

- It allows FVCs to choose a corporate governance regime that permits optimal structuring of the venture capital investment.
- It allows FVCs to avoid numerous administrative rules and technical restrictions on foreign equity ownership and foreign exchange conversion.[127]
- It allows FVCs to avoid the additional regulatory burden that comes from establishing a China-domiciled FVC.
- It allows FVCs to incorporate the holding company in a regime that provides more favorable exit routes than those that exist in China. The 'exit' issue will be discussed in detail in section 5 below. For now, it is worth pointing out that China has only recently provided a viable IPO strategy for its start-ups.

Chinese start-ups that benefit from this offshore strategy are frequently referred to as 'red-chip' companies. The term is a take-off from the U.S. term 'blue-chip' company, which refers to large, financially-sound companies with publicly-traded common stock.[128] The term red-chip company is used to describe publicly-traded offshore companies whose primary assets and business operations are located in China.[129] Finishing with a red-chip company is frequently the goal of the offshore strategy, which has resulted in the term 'red-chip' also being associated with the overall offshore FVC strategy. For the purposes of our analysis, and to distinguish the offshore

FVC strategy from other offshore strategies, we will refer to the above offshore FVC strategy as the 'red-chip' strategy.

3.3.2 China's regulation of the offshore strategy – round-tripping v. red chips

Until 2005, there was little regulatory oversight of the offshore strategy, and it was treated as 'legal' by lawyers and financiers based on the belief that 'whatever is not against the existing Chinese laws and regulations . . . should be allowed.'[130] The passive permitting of the offshore strategy came to an end, however, in January 2005.

While numerous FVCs and Chinese start-ups used the red-chip strategy to finance valuable, high-technology growth companies more efficiently, not every user of the strategy was employing it in such a socially beneficial manner. The offshore strategy was also being employed by a substantial number of China-domiciled investors to invest in Chinese companies and thereby transform the companies into foreign-invested entities. A purely Chinese company might decide to establish a Hong Kong, or Cayman Islands, holding company that is owned by the same China-domiciled investors that originally owned the Chinese company. The direct investment in the Chinese entity is thereby replaced with an offshore intermediary investment, and the Chinese company is now 'foreign-owned.' This approach – China-domiciled investors investing in a China-domiciled company through an offshore structure – is generally referred as 'round-trip investing' or 'round-tripping,' since the investment makes a round trip through a foreign holding company. Such round-trip investments could have socially-beneficial motives akin to those of the red-chip strategy, such as allowing Chinese companies to opt into more efficient regulatory regimes for part of their operations. At the same time, there is the distinct possibility that round-tripping could be used for less desirable reasons such as tax avoidance (until recently, foreign-invested companies benefited from a lower income-tax rate), money laundering or other undesirable behavior.

Not surprisingly, the offshore strategy eventually drew the attention of Chinese regulators. In early 2005, the State Administration of Foreign Exchange (SAFE) issued two circulars that brought the offshore strategy under the Chinese regulatory umbrella:

- On January 24, 2005, SAFE issued Circular No. 11: Circular on Issues Related to Improving the Administration of Foreign Exchange in Mergers and Acquisitions by Foreign Investors; and
- On April 8, 2005, SAFE issued Circular No. 29: Circular on Issues Related to Registration of Offshore Investment by Domestic

Individual Residents and the Foreign Exchange Registration of Mergers and Acquisitions Involving Foreign Investors.

Collectively, Circulars 11 and 29 imposed a significant and costly regulatory framework on the offshore strategy. SAFE approval, for example, was required prior to launching an offshore structure, while deals structured prior to Circular No. 11 had to be registered with SAFE. One of the bigger problems with Circulars 11 and 29 was the lack of details on the criteria for approval or registration.[131] Circulars 11 and 29 were designed to curtail 'round-tripping,' but they also ended up considerably slowing the red-chip strategy – in terms both of the flow of FVC capital into China and the ability of Chinese companies to list on foreign stock markets as red-chip companies.[132] It does not appear that slowing the red-chip strategy was intended, so following significant lobbying pressure SAFE repealed the prior circulars and replaced them with Circular No. 75. SAFE adopted Circular No. 75, Circular on Issues Related to Financing through Offshore Special Purpose Vehicles by Domestic Residents and Round-trip Investment, on October 21, 2005, and it went into effect on November 1. About one year later, MOFCOM entered the fray with the Provisions on Foreign-Funded Mergers and Acquisitions of Domestic Enterprises (the M&A Provisions), which were adopted on August 8, 2006 and went into effect on September 8. Circular No. 75 and the M&A Provisions retain the core purpose of the prior circulars, which was to require government approval prior to employing the offshore strategy. On the positive side, however, Circular No. 75 and the M&A Provisions better define the approval process (e.g., clarifying the criteria for approval and the documents required) and permit pre-approvals that allow parties to clear the regulatory hurdle before incurring the expense of structuring the actual deal.

3.3.3   Will onshore investing strategies replace offshore strategies?

While widely considered to be an improvement over Circulars 11 and 29, the government approval process in Circular No. 75 and the M&A Provisions have nevertheless increased the regulatory costs of the red-chip strategy. What the increased regulatory costs will mean for the future activities of FVCs is not entirely clear. The increased regulatory costs do not appear to have significantly dissuaded FVCs from investing in China, as the level of venture capital investments in China has remained steady from 2006 through 2008. On the other hand, the increased costs did not cause FVCs to immediately bring their operations onshore and operate as China-domiciled FVCs, although that could be changing. As more FVCs become comfortable operating under the regulations for China-domiciled venture capital enterprises – which lowers the regulatory cost for such regulations – and the domestic venture capital infrastructure continues to improve, it would not be surprising that a domestic strategy could become more cost-effective for FVCs than the traditional red-chip strategy.

3.4 Convertible Preferred Stock

3.4.1 Critical to venture capital investing strategies

Convertible preferred stock, or its equivalent, is critical to the success of venture capital investing strategies. Purchasing an ownership stake in a company typically entitles an investor to three fundamental rights:

1. A right to current distributions of the company's accumulated profits (e.g., dividends);
2. A right to the company's residual assets (e.g., the assets that remain upon the liquidation of the company after all of the company's liabilities have been satisfied); and
3. A right to vote on directors (or their equivalent) as well as certain extraordinary transactions (e.g., business combinations, amendments to the company's charter documents or dissolution of the company).

By specifically tailoring those three rights to address the particular information and management problems that plague start-ups, convertible preferred stock can greatly reduce the problems. For example, preferred stock has a higher priority than common stock, which means that holders of preferred stock get paid before holders of common stock if the company is liquidated.[133] Because the company's founding entrepreneurs and its management typically own common stock, preferred stock provides a substantial incentive to the managers to run the company successfully.[134] If they do not, the only parties who are likely to see any return on their investment will be the venture capital firms.[135]

Convertible preferred stock can also be used to provide venture capital firms with a number of contractual rights that better align the incentives of the start-up's managers with those of the venture capital firm, including:[136]

- rights to disproportionate representation on the board of directors;
- preferential rights to a disproportionate amount of the start-up's accumulated profits (e.g., mandatory dividends) and assets upon liquidation (e.g., participation rights);[137]
- mandatory redemption rights;[138]
- special voting rights;[139] and
- anti-dilution rights.[140]

Without going into detail on these various contractual rights, their general purpose is to provide the venture capital firm with substantial control over the start-up's assets and decision making, which protects the venture capital firm from the downside risk that accompanies venture capital investing as well as from potentially poor decisions that could be made by the start-up's managers.

The 'convertible' nature of convertible preferred stock is also very important, as it allows for an efficient transfer of this extensive venture capital control back to the start-ups managers if the managers demonstrate they have earned it through strong performance. Convertible preferred stock will almost invariably be designed to convert into common stock and thereby eliminate the venture capital firm's preferential contractual rights. While the typical venture capital convertible preferred stock will include both voluntary conversion by the venture capital firm and automatic conversion, it is the automatic conversion that is the most relevant for this analysis. The most common automatic conversion provision stems from the start-up conducting an initial public offering (IPO) and listing its stock on an attractive public stock market (e.g., Nasdaq's Global Market). In that situation, the start-up's managers have demonstrated their ability to run the company successfully and the venture capital firms have the ability to exit their investment (e.g., sell their stock through the public trading markets). While automatic conversions in the event of an IPO are the most common means for a venture capital firm to cede control back to the start-up's managers, it is not the only viable method for doing so. For example, the parties could contractually negotiate for a diminution in contract rights (or full conversion) in the event the start-up reaches certain financial metrics.

All told, the ceding of extensive rights to the venture capital firms coupled with their ability to earn those rights back through strong performance provides an efficient alignment of interest between venture capital firms and the start-up's managers. The start-up's managers are highly motivated to run the company successfully in order to regain control. Presumably, controlling their own destiny was a major factor in the founders'/managers' decisions to forego more secure jobs in order to start ventures with very high potential failure rates.[141]

3.4.2 Convertible preferred stock in China

Until recently, the ability to purchase convertible preferred stock was a major motivator for the offshore investing structure. China's corporate law did not formally recognize the validity of convertible preferred stock until 2006. Prior to that time, China's Company Law (1993) recognized ordinary shares as the only type of equity ownership in a Chinese company, which made the validity of convertible preferred stock issued by a Chinese company highly questionable.[142] Purchasing convertible preferred stock in a Chinese company prior to 2006, therefore, involved a significant degree of 'legal risk.' Generally speaking, venture capital managers are comfortable investing in a high-risk environment. Specifically, they are comfortable with the need to evaluate and manage technology risk (i.e., does the technology really work?) and managerial risk (i.e., is the management team capable of successfully running the company?) when investing in start-ups. While venture capital managers are comfortable with some types of risk, they are not comfortable with all types of risks, and legal risk is one type of risk they generally want to avoid. Venture capital managers are experts at evaluating technology and managerial risk, not legal risk. The offshore investing structure allowed venture capital firms to avoid the legal risk associated with Chinese convertible preferred stock by establishing the holding company in a jurisdiction that clearly permits convertible preferred stock. The venture capital firms do not invest in the Chinese operating company. Instead, they make their investment at the foreign holding company level (see Figure 8.2), where there is no question about the legal status of convertible preferred stock.

In March 2006, China's Provisional Measures for the Administration of Venture Capital Enterprises (2006 VCE Measures) came into effect which, among other things, authorized China-domiciled venture capital firms, including China-domiciled FVCs, to purchase preferred stock or convertible preferred stock when investing in unlisted, Chinese companies. Specifically, article 15 of the 2006 VCE Measures provides:

> Venture capital firms may purchase preferred stock, convertible preferred stock or other quasi-equity securities when investing in unlisted companies.

The 2006 VCE Measures followed the government's major overhaul of the Company Law (2005), which was also necessary to permit convertible preferred stock. Among a host of other changes, the revised Company Law clearly recognized the ability of Chinese joint stock limited companies (JSLCs) to issue multiple classes of equity ownership and to tailor the dividend, residual asset and voting rights that are so essential to a venture capital investing strategy (see Table 8.10).

The clear validity of convertible preferred stock for Chinese JSLCs may be another reason for the recently increased emphasis on China-domiciled VCFs. It should be pointed out that the 2006 VCE Measures apply only to China-domiciled venture capital firms, including China-domiciled FVCs, and do not apply to offshore VCFs.

Table 8.10   Select features of joint stock limited companies under the revised Company Law (2005)

Minimum registered capital	• RMB 5 million for domestic companies[1] • RMB 30 million for foreign-invested joint stock limited[2] companies
Permits multiple classes of equity ownership	The Company Law provides for multiple classes of stock for JSLCs[3]
Owners' rights to current distributions of accumulated profits	Profits are to be distributed in the manner set forth in the articles of association[4]
Owners' rights to residual assets upon liquidation	The company's residual assets, after all creditors have been paid, shall be distributed 'according to the proportions of stock held by the shareholders'[5]
Voting rights	Shareholders have one vote per share they hold.[6] The Company Law does not, however, prohibit different classes of stock from having particular 'class voting' rights

Notes:
1.   Article 81 of the Company Law.
2.   Article 81 of the Company Law provides:
      'If any law or administrative regulation prescribes a relatively higher minimum amount of registered capital, such provision shall be followed.'
      Article 7 of the Interim Provisions Concerning Some Issues on the Establishment of Joint Stock Limited Companies with Foreign Investment, promulgated by the Ministry of Foreign Trade and Economic Cooperation on January 10, 1995 provides:
      'Registered capital of a company [defined as a joint stock limited company with foreign investment] shall be the total capital stock recorded with registering departments. The registered capital of a company shall be at least RMB 30 million. Total value of the shares purchased and held by the foreign shareholders shall be no less than 25% of the company's total registered capital'.
3.   *See* Articles 129 and 134 of the Company Law.
4.   Article 82 of the Company Law.
5.   Article 187 of the Company Law.
6.   Article 104 of the Company Law.

4.   CREATING A VIBRANT VENTURE CAPITAL 'EXIT ENVIRONMENT'

The ability of venture capital firms to fulfill their financing and nurturing functions for the start-up market is highly dependent upon on the vibrancy of the venture capital exit environment. The exit environment impacts every aspect of the venture capital process[143] – including the availability of investment capital, the price venture capital firms are willing to pay, the stage at which venture capital firms will invest and the quality of the non-financial services the venture capital firms provide to their portfolio companies. This section will explain generally why vibrant venture capital/start-up industries are dependent on a healthy exit environment and will examine specifically the exit environment for venture capital firms operating in China. In particular, this section will examine China's new growth enterprise market – ChiNext – and its probable impact on China's venture capital/start-up industries.

4.1 Capital Gains Drive Venture Capital Firms

Having a clear path to liquidity is a key element of a venture capital manager's decision to invest, as most venture capital firms will not 'enter' an investment without a clear view of how to 'exit' it.[144] Venture capital firms do not represent a permanent financing option for start-ups, but instead represent what can be thought of as a 'medium-term' option.[145] While venture capital firms tend not to be short-term speculators, neither do they seek to become permanent owners of start-ups. Instead, venture capital firms seek to invest in start-ups with promise and then help nurture those start-ups until they have achieved a sufficient size and maturity to be attractive to a strategic acquirer or some other source of financing.[146]

Unlike many other forms of equity investments, start-ups are unlikely to generate cash payments for their investors absent an exit. In general, equity investments provide two primary methods for investors to generate cash: (1) dividends paid by the company from excess cash flow; or (2) capital gains earned from selling the investment. For equity investments in start-ups, the dividend method typically is not available. Since start-ups must generally sacrifice near-term profitability in order to build the company for growth, most start-ups do not generate positive cash flows[147] and therefore are not able to pay dividends.[148] As a result, the only viable way to generate cash from an equity investment in a start-up is to exit the investment and collect the capital gains.

4.2 Benefits to Venture Capital Firms and Start-ups of an 'Open' Exit Environment

Because of the importance of capital gains to venture capital firms, the 'openness' of the exit environment substantially impacts the various elements of the venture capital process. For example, an 'open' exit environment can lower the cost of capital for start-ups, increase venture capital activity and increase the allocation of investment capital to the most skilled venture capital managers.

4.2.1 Lowers the cost of capital for start-ups

The ability of venture capital firms to exit their investments has a substantial impact on the cost of capital they charge to the start-ups. Investment capital is a scarce resource relative to its need, and that scarcity demands a price.[149] As with any scarce resource, there are a variety of factors that influence the price for capital, including the ability for an investor to exit the investment. The ability to immediately exit an investment has value to investors and will factor into the price they are willing to pay for it.[150] Assuming two investments that are similar in all respects except for the ability to resell the investments, the investment with more impediments to its resale will trade at a discount to the investment that can be freely sold, as investors will treat the investment that is not immediately marketable as bearing an additional 'cost.'[151] This cost is meant to compensate the investor for the possibility that she may not be able to exit the investment when she believes it has reached its optimal value.

Because start-up investments can be difficult to resell – e.g., there is generally not a ready secondary market for such investments until the start-up is extremely mature, and the regulatory costs for exiting start-up investments can be very high – venture capital firms demand compensation for this reduced marketability in the form of a 'discount on lack of marketability.' Venture capital firms, therefore, discount the value they ascribe to the start-up by some amount that is meant to compensate for their inability to freely exit the investment. The more restricted the exit environment, the greater the discount (i.e., the greater the cost of capital charged to the start-up). This increased cost of capital may render private equity too expensive for a start-up and thereby cause the start-up to take a suboptimal amount of capital or forego taking the capital altogether. At a minimum, some amount of money that could have gone to funding the start-up goes to compensating venture capital firms for the negative impacts of reduced marketability. A more open exit environment will reduce the discount and decrease the cost of capital charged to start-ups.

4.2.2 Increases venture capital activity

As we explained earlier, most venture capital firms will not 'enter' an investment without a clear view of how to 'exit' it. Because of the importance of exits, venture capital managers will typically conduct due diligence on future exit scenarios as part of their initial decision to invest in the start-up.[152] If the start-up appears unlikely to be a candidate for an available exit strategy, venture capital managers are unlikely to invest in that start-up irrespective of the overall quality of the start-up.[153] An 'open' exit environment makes it easier for venture capital mangers to foresee the eventual exit, which greatly facilitates their investment decision.

4.2.3 Increases the allocation of capital to the best venture capital managers

Investment capital is not the only valuable, scarce resource in the start-up investment process. Competent venture capital managers who are skilled at identifying worthy start-ups and nurturing them as they grow are also in limited supply. The ability of these venture capital managers varies widely.[154] Exits provide the investors in a venture capital firm with a concrete opportunity to evaluate the fund managers' abilities. Many venture capital firms, including those that operate in China,[155] are structured as self-liquidating funds,[156] which means the venture capital firm's investment capital is invested only once, with any returns distributed immediately to the venture capital firm investors rather than being reinvested in the venture capital firm itself.[157] As a result, venture capital managers who wish to continue in the venture capital business (which appears to describe most such managers) must raise additional capital and create a further fund. Since exits are what determine the returns on investment, they provide investors with a measure of the ability of the venture capital managers and provide valuable information regarding whether additional funds should be allocated to those managers (if they are strong) or directed elsewhere (if venture capital managers are weak). In a more open exit environment, one should expect more exits to take place, which will provide investors with better information about the performance of the various venture capital managers.

4.3 Exiting Venture Capital Investments in China

There are two primary techniques that venture capital firms will employ to exit their start-up investments:

1. *Acquisition exits*, which involve the sale of the entire start-up to a third party. The third party acquires all (or substantially all) of the start-up in exchange for cash or stock of the acquiring company.
2. *IPO exits*, which involve the start-up conducting an IPO and creating a public secondary market for its stock on a liquid stock market. The venture capital firms can then sell their stock in the start-up through the secondary market.

Venture capital firms operating in China are no different from their counterparts around the world and they employ these same two techniques. Table 8.11 provides summary data on the number and dollar amounts of the two principal exit routes in China for the last ten years.

Since 2005, the exit environment for venture capital investments in China

*Table 8.11   Venture capital/PE exits from Chinese investments
                (1999–2008)*

Year	Venture capital/PE-backed acquisition exits			Venture capital/PE-backed IPO exits		
	Total	Total with disclosed_ values	Total value of disclosed acquisition exits (US$ millions)	Total	Total IPO offer amount (US$ millions)	Average size of IPO (US$ millions)
2008	21	10	1274.3	36	4848.4	134.7
2007	52	39	8377.9	109	48 605.2	445.9
2006	42	28	1773.2	50	57 313.3	1146.3
2005	35	18	2144.6	38	11 386.5	299.6
2004	13	7	286.3	38	7930.3	208.7
2003	5	4	345.8	18	6844.7	380.3
2002	3	2	78.1	6	461.4	76.9
2001	2	1	32.1	4	1282.7	320.7
2000	2	2	65.6	6	531.6	88.6
1999	0	n/a	n/a	5	169.1	33.8

Source: Zero2IPO Database

*Table 8.12   Relatively robust exit environment for venture capital/PE
                investments has developed in China (2005–2008)*

Period	Average annual no. of acquisition or IPO exits	Average annual value of disclosed VC/PE-backed acquisition exits (US$ millions)	Average annual value of VC/PE-backed IPO Exits (US$ millions)	Average annual combined total (US$ millions)
2005–2008	96	3392.5	30 538.4	33 930.9

Source: Zero2IPO Database

has shown itself to be quite robust (see Tables 8.11 and 8.12). The exit environment has recently demonstrated weakness, however, as 2008 showed a steep drop in both the number and value of Chinese acquisition and IPO exits compared to 2007. This trend has continued through the first half of 2009.[158]

4.3.1 How critical is a robust IPO market for start-ups?

There should be no doubt that successful venture capital markets require a robust exit environment that allows the venture capitalists to liquidate their investments. Many argue, however, that the availability of exit routes tells only part of the story. The *form* of available exits is just as critical to the development of a successful venture capital market.[159] Specifically, there is a commonly held belief that a vibrant venture capital market requires a well-developed stock market for start-ups that allows venture capital firms and their portfolio companies to reasonably project IPOs as the eventual exit strategy.[160]

The basic argument for the importance of IPO exits over acquisition exits revolves around two primary issues: IPO exits generate greater returns and IPO exits create a valuable 'control incentive.'

a. IPO exits generate greater returns   There is a widely-held belief in venture capital circles that IPO exits generate greater returns for venture capital investors than do acquisition exits. While there may be reasons to question the validity of such belief,[161] few in the venture capital industry seem to question the proposition that IPO exits are more profitable then acquisition exits. Greater returns provide a greater profit incentive for investors to invest in venture capital firms and for venture capital firms to invest in start-ups. In the United States, studies have shown that 'hot' IPO markets – i.e., periods when valuations are particularly favorable, and therefore even more profitable – are accompanied by a substantial increase in venture capital activity.[162]

On a related note, the potential of an IPO exit can also help to improve the bargaining position of start-ups that are considering acquisition exits. It is not uncommon for a start-up seeking an acquisition exit to find there are only a few potential companies that could be viable purchasers. The potential corporate buyers can exploit the start-up's lack of alternatives and acquire the start-up at an unfairly low price.[163] Having a viable alternative to acquisition exits, even if not employed, can help to provide the start-up with valuable negotiating leverage to obtain a more attractive valuation.[164]

b. IPO exits create a valuable 'control incentive'   Stanford Law School Professors Bernard Black and Ronald Gilson offer an alternative explanation for why acquisition exits alone are not sufficient for vibrant VC-firm/start-up industries.[165] In a leading article on the topic, Black and Gilson examined why the United States has both an active venture capital industry and IPO exits as a viable exit route for venture capital firms, while countries such as Japan and Germany have neither. Black and Gilson explain that the correlation is a result of what we will refer to as the 'control incentive.' In general, an acquisition exit will require the entrepreneur for the start-up to cede control of the start-up to the strategic acquirer, which is something that many entrepreneurs do not favor. Black and Gilson explain that a realistic potential for an IPO exit, even if the start-up ends up conducting an acquisition exit, is critical to an active venture capital market because of this control issue. The realistic potential for an IPO exit allows the entrepreneur and the venture capital firm to form an implicit contract over future control of the start-up.[166] Their theory is based on three assumptions: '(i) the entrepreneur places substantial private value on control over the company she starts; it is not feasible for a successful entrepreneur to retain control at the time of the initial venture capital financing; and (iii) it is feasible for a successful entrepreneur to reaquire control from the venture capitalist when the venture capitalist exits.'[167]

While the entrepreneur is required to give up substantial control over the management of the start-up when venture capital firms invest, the entrepreneurs likely wish to get that control back. Presumably, controlling their own destiny was a major factor in causing many entrepreneurs to leave secure jobs in order to start ventures with very high potential failure rates.[168] Black and Gilson posit that an IPO exit allows venture capital firms to exit at a time when the entrepreneurs have demonstrated an ability to run the start-up,[169] and control is returned to them as public investors are likely to be substantially more passive than the VC firms. This 'opportunity to regain control also provides an incentive, beyond mere wealth, for the entrepreneur to devote the effort needed for success.'[170]

4.3.2   ChiNext: China's new growth enterprise market

The perceived importance of IPO exits to a vibrant venture capital industry has led numerous countries around the world to create specialized stock markets – frequently referred to as growth enterprise markets (GEMs) – to serve the smaller, high-technology companies that venture capital firms finance and develop. China is no exception. In 2009, the Shenzhen Stock Exchange launched China's first GEM, which has been named ChiNext.[171] ChiNext has had an impressive start so far. The first 28 companies were listed on October 30, 2009, and they raised a total of 15.5 billion RMB that day.[172] Within three months of this launch, 50 companies had been listed on ChiNext with a total market capitalization approaching 200 billion RMB.[173]

Like GEMs in other countries, one of ChiNext's more distinctive features is the lower thresholds it sets for companies seeking to conduct an IPO. When compared to China's primary boards on the the Shanghai and

Shenzhen Stock Exchanges, ChiNext has much lower listing requirements regarding a company's age, historical profitability, size and public float. The Interim Measures on the Administration of IPOs and Listings of Shares on the ChiNext lay out the minimum thresholds (see Table 8.13).

4.3.3 The historical track record of GEMS is questionable – China should handle ChiNext with care

When creating a GEM, policymakers must inevitably strike a balance between two competing regulatory interests. On the one hand, there is a desire to minimize the regulatory burden of the smaller companies that are the targets of GEMs. As a general rule, regulatory compliance is disproportionately more expensive for smaller companies (e.g., as a percentage of profits or revenues) than for larger companies.[174] Expensive regulatory burdens that might be nothing more than a nuisance to larger companies could mean the difference between profitability and bankruptcy for a smaller company. So, if the regulatory burden is set too high, the smaller, high-growth companies that GEMs seek to service might be blocked from listing on the GEM.

On the other hand, the high-risk nature of smaller, public companies may warrant regulating them more heavily. While there is some debate regarding the ultimate goal for regulating publicly-traded companies, two of the more critical goals are:

* To reduce market problems (such as information asymmetries and agency problems) that expose members of the public to inappropriate risk; and
* To help to reduce the impact of potential 'lemons problems' (see Box 8.2) that are inherent to securities markets by signaling minimum standards of quality.

Smaller public companies are particularly prone to both lemons and investor protection problems. The greater risk of lemons and investor protection problems stems from certain characteristics that are common to a substantial portion of smaller public companies:

* They are riskier investments;
* They are more prone to making inaccurate financial disclosure,[175] which is the most fundamental information that a public company communicates to its investors;
* They are less likely to receive meaningful support from valuable securities intermediaries, such as research analysts, that help to improve the efficiency of their stock prices;[176] and

Table 8.13   Select minimum listing requirements for ChiNext

Requirement	Description
Corporate Form	Company must be a JSLC[1]
Age	Must have a continuous operating history of at least three years[2]
Profit History	Company must have been either: • Profitable in each of the two prior fiscal years and have accumulated net profits of at least RMB 10 million. The profits must be growing steadily;[3] or • Profitable in the most recent fiscal year with revenues of at least RMB 50 million and a net profit of at least RMB 5 million that year. The company's revenue growth rate for the last two years must be at least 30%.[4]
Net Assets	Pre-IPO net assets of at least RMB 20 million.[5]
Share Capital	Post-IPO share capital of at least RMB 30 million.[6]
Continued Profitability	Company must demonstrate that it will continue to operate profitably by showing, among other things: • Its industry has not experienced a material adverse change and it not likely to do so in the future; • Its position in the industry has not experienced a material adverse change and it not likely to do so in the future; and • There is no material risk regarding the intellectual property rights it needs to conduct its business.[7]
Company Stability	In the two prior fiscal years, there has not been a significant change in the company's 'principal business, directors and senior management . . . nor any change of its *de facto* controller.'[8]

Notes:
1. Article 10(1), Interim Measures on the Administration of IPOs and Listings of Shares on the ChiNext.
2. *Id.*
3. Article 10(2), Interim Measures on the Administration of IPOs and Listings of Shares on the ChiNext.
4. *Id.*
5. Article 10(3), Interim Measures on the Administration of IPOs and Listings of Shares on the ChiNext.
6. Article 10(4), Interim Measures on the Administration of IPOs and Listings of Shares on the ChiNext.
7. Article 14, Interim Measures on the Administration of IPOs and Listings of Shares on the ChiNext.
8. Article 13, Interim Measures on the Administration of IPOs and Listings of Shares on the ChiNext.

BOX 8.2 LEMONS PROBLEMS

Where asymmetric information prevents investors from being able to distinguish good companies from bad companies, a classic 'lemons problem' occurs.[1] In such a setting, investors are likely to view all investment opportunities as roughly average. This means that investors will likely underpay for good companies (i.e., insufficient capital is dedicated to good companies, since they are judged as average) and overpay for bad companies (i.e., too much capital is allocated to them, since they are also judged to be average). Because good companies are penalized by this effect, they will strive to differentiate themselves from bad and average companies. If good companies cannot differentiate themselves, they will likely leave the market as they will be disappointed with the price paid by investors. These good companies will seek alternative forms of financing that provide them a more appropriate cost of capital. Left unchecked, this lemons problem will cause bad companies to dominate the market because they will be the most motivated by the average price paid by investors – jeopardizing the very existence of the market.[2]

Notes:

1. The impact of asymmetric information on markets and the resulting 'lemons problem' can be traced back to George A. Akerlof (1970), 'The Market for "Lemons": Quality Uncertainty and the Market Mechanism', 84 *Q. J. Econ.* 488, 490–492. Mr. Akerloff won the Nobel Prize in Economics for this 13-page paper.
2. *See generally id.*

● Their primary public shareholders are more likely to be ordinary, retail investors (see Box 8.3) whose overall level of investment sophistication is generally quite low.[177]

These characteristics for smaller public companies make GEMs particularly ripe for boom/bust scenarios. If policymakers lower the regulatory cost for accessing the public markets through a GEM, but do not develop regulatory institutions that adequately address the greater risks posed by smaller public companies, the GEM can encourage an inappropriate risk transfer of higher-risk investments from investors that are specialized in such higher-risk investments (e.g., venture capital funds) to less

---

BOX 8.3   INSTITUTIONAL V. RETAIL INVESTORS

Investors in public stock markets are not a homogenous group. Different investors have varied levels of sophistication and ability to intelligently evaluate the risk of an investment and ascribe a reasonable value to it. For illustrative purposes, we will oversimplify the matter and group investors into two broad categories: institutional investors and retail investors.

- Institutional investors, such as mutual funds, insurance companies, pension funds, investment banks, or other entities which regularly invest large amounts of money in the securities markets are generally sophisticated investors. They have the ability to intelligently investigate and value securities investments and they appreciate when they are incurring heightened risk with an investment.
- Retail investors are basically everyone other than institutional investors. Retail investors have historically demonstrated an inability to intelligently investigate and value securities investments. Moreover, they tend not to appreciate how unsophisticated they are as investors or realize how much risk they are truly incurring when investing.

Source:

John L. Orcutt (2009), 'The Case Against Exempting Swaller Reporting Companies from Sarloqnes–Oxley Section 464: Why Market-Based Solutions are Likely to Harm Ordinary Investors', 14 *Fordham J. Corp. & Fin. G.* 325, 405–406.

---

sophisticated retail investors. Unfortunately, most GEMs appear to have fallen into this trap.

Consider the following potential scenario. A GEM opens with reduced regulatory requirements, coupled with a lot of promotional publicity (e.g., announcements by government officials, business leaders and the media) about the value of the GEM and its ability to finance the next generation of great technology companies. The GEM, however, has not developed tailored regulations to prevent an inappropriate risk transfer to unsophisticated retail investors. The first companies listed on the GEM are carefully selected, successful companies, so the GEM gets off to a successful start with strong company performance and even more positive media attention. The success of the first wave of companies draws the attention of both the general public and other smaller, technology companies, not all of which are as strong as the first wave of companies. The influx of unsophisticated retail investors has an inflationary impact on the GEM's stock prices. New issuers, attracted by the generous stock valuations and taking advantage of the reduced listing requirements, also come into the market and lower the quality of companies on the GEM. Fraud, scandals and disappointing results follow, which motivates the best companies to leave the GEM and seek higher prestige markets, including going overseas to list on Nasdaq, which further lowers company quality on the market. Institutional investors leave the market, which reduces the market's liquidity and starts to drive down stock prices, and within a relatively short period of time an extreme bust environment sets in and the market risks irrelevance or closure.

While we have not done a market-by-market analysis of all GEMs, it is safe to say that GEMs round the world have suffered from an inordinate number of busts and outright closures during the last decade or so. Consider the fates of Belgium's EASDAQ, France's Nouveau Marché, Germany's Neuer Markt, India's Indonext, Italy's Nuovo Mercato, and Switzerland's SWX New Market – to name just a few. Each of these markets experienced extreme busts and either closed or was reorganized into a new entity for a fresh start. Chinese policymakers need to exercise extreme caution with ChiNext. Chinese officials are aware of the checkered past of most GEMs, and have tried to develop ChiNext in such a way as to avoid the extreme boom/bust problem. Whether they have been able to find the right mix of reduced regulatory cost v. stronger regulations to ward against an inappropriate risk transfer to the general public is not yet clear.

5. THERE IS PLENTY OF REASON TO BE OPTIMISTIC ABOUT THE FUTURE OF VENTURE CAPITALISM IN CHINA

Despite our reservations about the future success of ChiNext, there is plenty of reason to be optimistic about China's venture capital market and its ability to finance and support a Chinese start-up industry. The growth in venture capital activity – in terms both of the amount of venture capital investments and the sophistication of the venture capital managers – is undeniable. China is truly developing a modern, sophisticated venture capital industry. The evolution of China's legal system to support such a modern venture capital industry gives us lots of confidence in the future of venture capitalism in China.

We would like to offer, however, one strong cautionary note. China's venture capital industry is still young and it has yet to face challenges that are certain to present themselves as the industry matures. One such challenge stems from the cyclical nature of venture capitalism. Venture capital industries are highly prone to both 'boom' and 'bust' periods. During such boom/bust periods, venture capital firms have a tendency to behave in a number of socially undesirable ways. In boom periods, for example, venture capital firms have shown a tendency to try and prematurely foist overly-risky companies on an unsuspecting general public. In developing a regulatory/policy environment that supports venture capitalism, Chinese policymakers need to keep in mind that boom/busts will occur in the future – as far we can tell, they are inevitable – and that the general public should be shielded, to the extent possible, from the dangers of such boom/busts.

PART III

The future

9. Increasing the technology commercialization capacity of Chinese universities

> Universities in China are yet to become the key driver of national research and development.
>
> – Weiping Wu[1]

The last decade has been a good one in China for developing university technology commercialization capacity.[2] Chinese universities have greatly improved their ability to create and diffuse commercially-useful technology. University patent activity and licensing activity have consistently and dramatically increased – in terms of both quantity and quality. University-created S&T start-ups have seen similar progress and universities continue to commercialize meaningful amounts of technology through non-patent technology contracts. Despite these remarkable improvements, university technology transfer remains at the periphery of China's innovation system. China's university technology transfer system is not yet at a place where Chinese policymakers can count on its universities to consistently provide the innovation and technological advances that Chinese companies will need to be competitive in a knowledge-based economy.[3] Considering the resources that China commits to its universities and the policy emphasis it has placed on university technology transfer, it is not unrealistic for policymakers to expect university technology commercialization to move in from the periphery and become central to the country's technology-based economic development movement. Chinese universities 'should' become substantial drivers of economic growth and well-being.

In order to achieve this ideal outcome, however, China's university technology transfer system will need to grow. Considering the relative youth of the system, the performance has been quite impressive. In fact, it is realistic to project that university technology transfer will continue to grow in the near term merely from the various actors in the system gaining valuable technology transfer experience. Having said that, significant weaknesses do exist in China's university technology transfer system that will impede growth and reduce the system's overall efficiency. We see four issues in particular that, if resolved, could go a long way towards strengthening the technology commercialization capacity of Chinese universities:

1. Improve the environment for technology transfer offices in Chinese universities.
2. Look beyond Silicon Valley and Route 128 for valuable lessons on improving university technology commercialization.
3. Continue to improve the environment for high-technology start-up companies in China – in particular the environment for the earliest-stage start-ups.
4. Increase awareness of the role that university technology transfer could play in China's economy.

As with any complex system, there are of course many, many more weaknesses that could be considered. In highlighting these four issues in particular, we have tried to focus on non-obvious issues that are also realistically resolvable. So, for example, we are not focusing on whether government R&D funding is sufficiently market-based in China. The short answer is that it probably is not,[4] but there is no reason to believe that Chinese policymakers are not already aware of the problem and understand the basic techniques for improving the situation. Ironically, while the primary purpose of this book has been to examine whether China's legal system is sufficient to support a vibrant university technology transfer system, we have not found China's laws, regulations or legal institutions to be among the major challenges for growing university technology transfer in China.

1. IMPROVE THE ENVIRONMENT FOR TECHNOLOGY TRANSFER OFFICES IN CHINESE UNIVERSITIES

China's implementation of a Bayh–Dole system substantially increased the economic incentives for Chinese universities to create and diffuse commercial technology. In order to take advantage of these increased incentives, universities needed to develop infrastructure to manage the process. Universities needed to organize an efficient interface between their research capabilities and industry. Beginning in the late 1990s, a significant number of research universities established internal technology transfer offices (TTOs) for commercializing university-developed technology.[5]

1.1 Traditional TTO Functions

In the United States, where the most developed TTOs tend to reside, the principal activities for TTOs can be roughly described as follows:

- *Identify the technology:* TTOs mine their university for valuable technology that should be patented (or copyrighted). This mining effort can take many forms, but frequently centers on developing routine disclosure systems to keep the TTO informed on technological innovations by university researchers.
- *Protect the technology:* once identified, TTOs will evaluate the commercial potential of the technology to determine whether it should be patented (or copyrighted). For those technologies that should be protected, the TTO will be responsible for obtaining that protection.
- *Market the technology:* TTOs are also responsible for developing and executing marketing strategies for the technology, such as identifying commercial licensors and negotiating licensing agreements with them.
- *Shape the university's technology transfer environment:* the best TTOs also play a more subtle role – they help to shape the university's technology transfer environment. The ability for universities to consistently commercialize technology is an inherently complex endeavor that requires coordinating an array of different activities and actors (e.g., university researchers, industry, university administrators, etc.). Effective linking of these various activities goes a long way to creating a successful commercialization environment. TTOs are the actors that are best positioned – by far – to link these various activities and actors and develop institutional practices that promote consistent technology commercialization.

Performing these functions (Traditional TTO Functions) helps to make a university's technology commercialization efforts more efficient. These Traditional TTO Functions help to reduce transaction costs and information asymmetries that would otherwise substantially increase the costs of technology commercialization. While often overlooked by policymakers, TTOs are absolutely critical to university technology commercialization. TTOs in China perform the Traditional TTO Functions, but they do so with varying degrees of competence.[6]

1.2 Brief History of Chinese TTOs

The earliest TTOs in China date back to 1985, following the launch of the eventually-failed technology market.[7] As we explained earlier, total university patent applications ranged from roughly 1000 to 1500 per year from 1985 to 1998, and TTOs did little to accelerate technology commercialization during this period.[8] At some universities, rather than rely on TTOs, they established their own IP law firms to provide any needed IP protection and patent application services.[9] Overall, TTOs were not overly significant technology actors during the 1985–1998 timeframe.

This lack of TTO capacity had a very negative impact on technology commercialization in China. At the most basic level, the lack of TTO capacity caused many university researchers to simply eliminate the university from any commercialization strategy by classifying their inventions as being not 'inventions for hire.' Without TTO capacity, universities were frequently unaware of, or unconcerned about, their potential rights to such inventions, and therefore failed to challenge whether the inventions should in fact be classified as inventions for hire. The practical effect was that university technology transfer in China during the 1990s very much resembled the Japanese experience of the 1990s.[10] Rather than following conscious university strategies for technology commercialization, most inventions transferred to the business sector were the result of individual professors patenting the invention on their own and individually negotiating the transfer to business enterprises. Kneller explains:

> Such an 'inventor retains rights' system allows for smooth and efficient transfers when there is a good match between the invention and the companies with which the professor/inventor happens to have longstanding relations, and where the technology has clear commercial potential. However, the Japanese experience shows that such a system provides weak incentives for either inventors, universities, or companies to develop inventions when the commercial value is not immediately obvious.[11]

Since the late 1990s, that situation has changed dramatically. With the government's increased focus on commercializing university research and the increase in university TTO capacity, universities in China now take a much more active role in asserting their rights over university inventions.[12] Many new TTOs were established,[13] while a number of universities with existing TTOs provided them with increased resources to commercialize technology.[14] For example, universities such as Peking University, Tsinghua University and Beijing University of Industrial Engineering established patent funds to support faculty patent applications.[15] In 2001, the Ministry of Education required the university-affiliated law firms to separate from the universities and become independent legal entities.[16]

The increased emphasis on TTOs in the late 1990s corresponds to the government's strong regulatory push to increase university technology commercialization. TTOs are now recognized as a necessary part of university technology transfer, but the development of their value-added role in the innovation process is still relatively new.

1.3 Competence of Chinese TTOs – The 'Isolation Problem'

The Traditional TTO Functions are highly sophisticated tasks that require a substantial amount of expertise. Moreover, much of that expertise comes from experience at doing the tasks over a number of years. While this is a bit of an over-simplification, most of China's TTOs did not begin to truly develop their sophistication for performing the Traditional TTO Functions until the late 1990s – and more specifically 1999. As a result, China's TTOs have faced a steep learning curve over a relatively short time frame.

Consider just a few of the questions that TTOs and their universities need to resolve in order for the TTO to have a reasonable opportunity to perform its functions effectively:

- What is the most effective way to organize a TTO? Should the TTO be an independent office that is solely responsible for IP management, or should the TTO be part of the university's office of sponsored research (which focuses on applying for government grants, supervising the resulting projects and fulfilling the reporting requirements for such grants)?
- What level of resources should be dedicated to the TTO?
- What should be the university's institutional IP policies? What methods should the university use for identifying valuable potential IP? How should the university define inventions for hire? How does the university ensure that its researchers understand the commercialization process and the requirements for maximizing the university's IP rights?
- What are the ideal policies for revenue sharing between the university and the researcher on inventions?
- How should the university manage conflicts of interest that tend to arise between a researcher's obligation to the university and its obligation to industry partners?
- What are the most effective techniques for marketing IP to industry?
- What are the challenges and pitfalls that come with creating start-ups?
- How do you value IP?
- How can the TTO institutionalize the knowledge that it obtains from its various commercialization efforts and experiences?

Some universities and their TTOs have made substantial progress at addressing these challenges and have developed highly effective TTOs. Tsinghua University's TTO, for example, is frequently held up as a model TTO. Not surprisingly, however, many other TTOs have struggled with any number of these challenging questions[17] and operate at a relatively low level of competence. A poorly functioning TTO *absolutely reduces the ability of a university to efficiently develop and commercialize useful technology*. Creating an environment that leads to better TTOs and better TTO performance should be at the top of any list for improving university technology transfer in China. As we will explain below, it is a relatively easy problem to fix, and yet the solution has the ability to dramatically improve technology transfer results.

One of the biggest reasons for the uneven competence of Chinese TTOs is the fact that they operate in relative isolation. Each TTO needs to go through the bumps and bruises of developing technology transfer best practices on its own. In order to achieve optimal competence, each TTO has to individually go through a very expensive learning process that universities may not be willing to pay for. The TTOs' solitary operations lead to what economists call a 'collective action problem.' Namely, while the cost of going through the learning process may be justified by the benefit to TTOs as a class, such cost is greater than the benefit that would be received by any one TTO.[18] For example, a single TTO that spends money to develop a database of licensing rates will learn a lot about IP valuation that could help that particular TTO. But if the cost of developing that database is prohibitively high, then the cost of developing that knowledge will not be justified by the benefit. If both the cost and benefit of the learning process can not be shared collectively across a large group of TTOs, it should be expected that such information learning processes will not occur, and TTOs will not be able to function at anywhere near an optimal level.

1.4 Fixing the 'Isolation Problem' – Establish an Association of Chinese TTOs

In the United States, the Association of University Technology Managers (AUTM) is employed as an effective tool for resolving the collective action problem and addressing many of the above questions that TTOs must address. Founded in 1974, AUTM seeks to assist university technology managers in a number of valuable ways, including:[19]

- *Collect and disseminate information:* one way to improve the operation of TTOs is to increase the information about university technology transfer – e.g., patent application and grant information, licensing information, university start-up information – which they have at their disposal. In addition to posing a collective action problem, collecting such information runs into a 'proprietary information' problem. Much of the information comes from individual TTOs which would probably not provide such information unless they were provided with access to similar information from other TTOs. AUTM serves as a valuable collectivizing agent for U.S. TTOs to overcome these problems. AUTM has developed the most comprehensive database on U.S. university technology transfer information, which it makes available to its members.

- *Training and professional development:* AUTM provides training and professional development services to its members to help them to understand best practices in university technology management and to develop standard contracts and procedures. These services help to increase the expertise of TTOs on a broad level. Rather than simply institutionalizing knowledge in one single TTO, AUTM helps to spread that knowledge broadly throughout a multitude of TTOs. Moreover, AUTM can again serve as a valuable collectivizing agent. By spreading the costs over a broad group of TTOs, AUTM can afford to invest more heavily in training, professional development and document development than can any one TTO.
- *Improved networking:* by holding various meetings, AUTM helps to link TTO professionals from an array of different universities so that they can form professional relationships. If a TTO at one university runs into a challenging situation, having professional colleagues at other TTOs that may have faced similar challenges can be an extremely valuable resource.

One challenge that is proving to be particularly difficult for Chinese TTOs is how to properly value the IP they seek to commercialize.[20] Universities seeking to commercialize more advanced technology are frequently faced with a thin technology buyer's market that may consist of only one multinational corporation as a buyer. In that setting, the multinational corporation will likely have much more experience negotiating deals on that type of technology. Such experience gives the buyer a significant advantage when negotiating against the TTO. What are the typical licensing terms for that type of licensing deal? How are such deals valued and structured? Having access to the collective knowledge of a broad group of TTOs that an organization such as AUTM provides could be very valuable for dealing with this type of valuation problem and putting the TTOs on a more even playing field with multinational corporations.

In China, unfortunately, there is no equivalent to AUTM, which we believe leads to slower development of expertise in Chinese TTOs, a lower competence in the Traditional TTO Functions and a significant negotiating disadvantage vis-à-vis industry technology buyers. Chinese universities and TTOs should seriously explore self-organizing an organization such as AUTM. Without such a collectivizing force, Chinese TTOs are each left to go through the TTO learning process on their own, which is a very time-consuming and costly process – in particular when multiplied over hundreds of different TTOs.

There are alternatives to establishing an association of Chinese TTOs. For example, the Chinese government could sponsor formal training for TTO personnel. Such training – which is not uncommon for developing technology transfer systems – could be offered by international experts and quickly inject Chinese TTOs with international best practices.

Ideally, Chinese TTOs would both receive formal training from international experts and establish an association of Chinese TTOs. The two solutions are not mutually exclusive, although we do believe the association would provide the more fundamental and long-term benefit to China's research universities. Training from international experts provides, at best, a partial solution for China. On the positive side, such international training provides an immediate infusion of knowledge and expertise. On the negative side, however, such training provides only a one-time solution to the collective action problem that constantly plagues TTOs. As practices change, how will China's TTOs continue to have access to evolving best practices? The association provides the long-term solution to that problem. Moreover, China's innovation system is an extremely complex creature with many elements that are truly unique to China. In order for international trainers to provide optimal TTO training, they need to fully understand China's innovation system. We are skeptical that most international trainers will have such in-depth knowledge. Again, the association provides a solution to this shortcoming, as the association would be well positioned to both (i) seek international best practices, and determine how those best practices would function in China's unique innovation system, and (ii) seek Chinese solutions to technology transfer issues that arise in China.

2. LOOK BEYOND SILICON VALLEY AND ROUTE 128 FOR VALUABLE LESSONS ON IMPROVING UNIVERSITY TECHNOLOGY COMMERCIALIZATION

2.1   The Silicon Valley/Route 128 Paradox

There should be little argument that Silicon Valley and the Route 128 region (the greater Boston area) provide ideal examples of the economic growth that can flow from consistent commercialization of university technology. Berkeley, M.I.T. and Stanford have been instrumental in supplying their economies with valuable commercial innovations around which a multitude of successful businesses – and even entire new industries – have been built. The success of Silicon Valley and the Route 128 region, however, creates a rather deceptive trap for the unwary. If policymakers, or university administrators, want to improve the technology commercialization capacity of a university, there is an intuitive appeal to simply study the examples of Berkeley, M.I.T. and Stanford and copy their best practices. Unfortunately, these three universities, despite their indisputable success, tend not to provide the best lessons for most universities.

The explanation for this apparent paradox is quite simple – few other universities are blessed with anything close to the favorable technology commercialization environment of a Silicon Valley or a Route 128 region. Consider just some of the advantages possessed by Berkeley, M.I.T. and Stanford:

- Some of the most talented university researchers in the world;
- Incredibly strong technology commercialization cultures;
- Two of the most fertile environments of high-technology companies (including both start-up and larger, well-established companies);
- Probably the strongest collection of intermediaries for encouraging the commercialization of university technology (e.g., venture capital firms, IP and start-up lawyers, entrepreneur support groups) that exists anywhere in the world; and
- A proven track-record of successful research and commercialization efforts that helps to attract more (i) government R&D funding; (ii) talented university researchers, (iii) high-technology companies and entrepreneurs that seek to acquire university technology, and (iv) intermediaries to facilitate the entire model.

With those advantages, it is no surprise that Berkeley, M.I.T. and Stanford are so successful. How many Chinese universities, however, are blessed with such advantages? To be blunt – none. But a handful of Chinese universities, led by Tsinghua University, are able to approximate some level of those advantages. Tsinghua, for example:

● Has some of China's most talented university researchers;
● Is located in one of China's most economically advanced regions that possesses a strong business sector;
● Beijing has one of the strongest collection of intermediaries in China; and
● Tsinghua has one of the strongest university brands in China, which helps Tsinghua to attract (i) a disproportionate share of Chinese R&D funding; (ii) the best researchers; (iii) interested high-technology companies and entrepreneurs; and (iv) intermediaries.

Only a few Chinese universities have comparable advantages, which helps to explain why effective technology commercialization has so far been limited to a relatively small number of Chinese universities (see Chapter 6, Section 2). These favored few Chinese universities can follow the Silicon Valley/Route 128 model – e.g., hire the best researchers, obtain a disproportionate share of R&D funding, develop strong relationships with the top high-technology companies that populate their region, and work with the various intermediaries in their region – and expect favorable results.

What about the rest of China's research universities that are not blessed with such advantages? What should they do to improve their technology commercialization capacity? They cannot simply decide to hire the best university researchers – since there is only a limited supply and the best have likely already been hired by the top handful of universities. They cannot simply decide to receive a greater amount of government R&D funding – and it might not be cost-effective for the government to just simply provide them with more R&D funding. They cannot simply decide to have more top high-technology companies and intermediaries in their region. To appreciate the magnitude of the problem for many of these universities, one needs to remember the large disparity in economic development of China's various provinces – i.e., 'one China, four worlds'[21] (see Chapter 5, Section 2.1.1).

2.2 Emulate More Realistic Models for Success

Should Chinese universities that are not blessed with the advantages of a Tsinghua simply give up trying to develop technology commercialization capacity – since their efforts are likely to be futile? Absolutely not.

Even if a Chinese university is not likely to ever become the next M.I.T. (or the next Tsinghua), universities that improve their technology commercialization capacity can still generate tangible benefits to themselves and their local economies that far exceed the costs of the efforts. The key is to seek successful models that are more suitable for their particular situations.

Remembering that commercialization of university technology takes place within a 'system' can be very helpful in finding a suitable model to emulate. Before blindly adopting a particular technology commercialization model that appears on its face to be successful, university administrators should conduct an in-depth analysis of their particular technology commercialization system – including the system's strengths and weaknesses. That way they can seek examples of successful technology commercialization universities with comparable profiles. In the United States, there is a rich diversity of universities that have been able to consistently generate considerable technology commercialization results while having to overcome significant system weaknesses.

Dartmouth College provides an excellent example of a successful commercialization effort that needed to overcome a major weakness. Dartmouth is located in a sparsely-populated, rural area of western New Hampshire that will never be confused with Silicon Valley or Route 128. The school is not surrounded by a dominant high-technology business sector, nor is it teeming with sophisticated technology transfer intermediaries. To top things off, Dartmouth is a relatively small school, which makes investing in R&D infrastructure more expensive for Dartmouth since it has fewer students and professors to justify the large fixed costs involved with building research labs. And yet, Dartmouth has been able to overcome these system weaknesses to create a successful technology cluster for itself and that benefits its surrounding community.[22]

Dartmouth's technology commercialization success has greatly benefited from focused improvement efforts by the school. One such effort was Dartmouth's establishment of the Dartmouth Entrepreneurial Network (DEN) in late 2000.[23] DEN works with researchers, alumni and students to support entrepreneurial initiatives by providing a variety of beneficial services. Much of the focus of those services is to provide entrepreneurs with invaluable mentorship and with access to a variety of contacts and services that would not otherwise be available to them. One of the most valuable elements of DEN is its ability to connect researchers and entrepreneurs with Dartmouth's powerful network of alumni.[24] Since January 2001, DEN has provided support for over 200 projects and companies.[25]

The important lesson to take from the Dartmouth example is not necessarily the specifics of the DEN model. The important lesson is that creative solutions can be found to improve technology commercializa- tion at universities that have serious disadvantages to overcome, and that many of those solutions have already been developed and tested. There is no single, perfect model for technology commercialization that will work in all settings, and Chinese universities should spend the time to find a model that suits their particular strengths and weaknesses.

2.3   Focus on Generating Incremental Benefits in a Cost-Effective Manner

In addition to emulating suitable models, universities also need to estab- lish reasonable expectations. For most universities, their technology com- mercialization efforts should focus on generating incremental benefits rather than on 'world-altering' benefits. Establishing realistic expectations helps universities to avoid foolish technology commercialization efforts that cause them to lose money and thereby unfairly view technology com- mercialization as a money-losing effort that should be avoided. One of the keys to building long-term technology commercialization capacity is to avoid such money-losing efforts.

The 'potential' benefits from technology commercialization can be heady stuff for policymakers and university administrators to consider. Consultants come in and start telling wonderful success stories – e.g., how universities like Columbia generated $135 million in licensing income in 2007 and M.I.T. generated $61 million[26] – and the dollar signs start to cloud people's expectations. Expectations can begin to exceed the realistic results that a university can generate, and too much money gets invested into the university's technology commercialization efforts. The university overspends on importing high-profile faculty members with the hope they will jumpstart the university's technology commercializa- tion success. The university builds expensive laboratories and incubators that are not justified by the university's technology efforts, or dramati- cally over-staffs its TTO. When the expensive professor does not work out, or the incubator turns out to be a bust, the technology commerciali- zation efforts looks like a failure because it falls below the universities' unreasonably inflated expectations and lots of university money has been wasted.

The better strategy, by far, is to develop realistic expectations and to focus on building a technology commercialization system that achieves a variety of benefits, beyond just licensing income, including:

- Creating a more effective mechanism for getting university research out of the laboratory and into the hands of the public – which is valuable even if the university is only generating a small amount of useful research.

- Creating a more attractive work environment that helps the university to attract better researchers.
- Increasing communications between the university's researchers and the business sector so that professors obtain first-hand knowledge of industry needs. Such knowledge can (i) help professors to design courses and curricula that are more responsive to industry and increase the university's ability to influence the placement of its students, and (ii) positively influence the direction of future R&D efforts.
- Providing the university with a more direct role in economic development, which can provide the university with a stronger voice when negotiating for greater access to government funding.

Once the university identifies the benefits that it can realistically achieve, it should then work to put a realistic dollar amount on the value of those benefits. That way, when the university develops its own particular technology commercialization plan, it can construct a cost-effective plan (i.e. its costs are in line with the realistic benefits that can be achieved).

3. DEVELOP A 'START-UP PIPELINE'

Chinese policymakers clearly appreciate the importance of a vibrant start-up industry. They also understand that creating a supportive financing environment for such companies is one of the most fundamental conditions for a vibrant start-up industry. Chapter 8 chronicled the impressive strides that China has made to promote venture capitalism in China. Unfortunately, those efforts are unlikely to provide a long-term solution for China's start-up industry unless something is done to help finance the earliest-stage start-ups. Venture capital firms tend to come in near the tail end of the start-up process. In order for venture capital firms to be relevant, China needs to create an environment that encourages the continuous creation and growth of a pipeline of start-ups that will eventually seek venture capital financing. China could do much more to develop such a pipeline.

3.1 Funding Gap for Early-Stage Start-ups

Most of the attention that surrounds start-up financing focuses on venture capital firms. Venture capital firms are typically portrayed as the primary

Table 9.1   Financing sources by funding stage

	Seed Stage	**Early Stage**	**Later Stage**
Primary financing sources	• Entrepreneurs themselves • Friends and family	• Angels	• Venture capital firms
Secondary financing sources	• Angels	• Early-stage venture capital firms • Strategic partnerships with larger companies	• Angels • Strategic partnerships with larger companies

Sources: John L. Orcutt (2005), 'Improving the Efficiency of the Angel Finance Market: A Proposal to Expand the Intermediary Role of Finders in the Private Capital Raising Setting', 37 *Ariz. St. L. Rev.* 864, 875; U.S. Gen. Accounting Office (2000), *Small Business Efforts to Facilitate Equity Capital Formation*, 9.

funding source for the start-up community.[27] In fact, venture capital firms play a much more limited role in financing start-ups. While this is a bit of an over-generalization, it is fair to say that venture capital firms generally restrict their investments to 'later-stage and larger deals.'[28] To illustrate this point, one can divide the financing needs for start-ups into three fundamental stages:[29]

1. *Seed stage:*   typically involves relatively small amounts of money that are used to get the company started (e.g., hire a few initial employees, secure office space and conduct product development) and to determine if the venture is worth pursuing.
2. *Early stage:*   if the start-up shows promise, the founders will seek to obtain the company's first external funding. Early-stage funding is used to complete product development, begin marketing, and commence an initial roll-out of the company's product or service.
3. *Later stage:*   as the company grows, so does its need for additional capital. Later stage funding is limited to those companies that have demonstrated some level of success and are looking to finance a major expansion.

Venture capital firms tend to focus predominantly on later stage funding and avoid earlier stages. Table 9.1 provides a summary of the typical financing sources for the three funding stages. Venture capital financing is only relevant for those start-ups that are able to obtain enough funding so that they can mature into later stage companies. Vibrant start-up industries require a financial bridge that takes start-ups from the seed-stage internal sources of funding to the later-stage deep pockets of the venture capital firms.

3.2 One Solution to the Funding Gap – Encourage the Development of a Robust Angel Finance Market

In the United States, this financial bridge is provided by angel investors. As we explained in Chapter 8, the term angel investor (or business angel) refers to wealthy individuals who invest their own capital directly in start-ups, rather than rely on a financial intermediary such as a professional venture capital manager. Because angel markets are made up of large numbers of individual investors from a multitude of backgrounds, it is difficult to characterize angels with much precision. Nevertheless, some useful generalizations can be made about angels and their role in financing start-ups:

- In the United States, at least, many are former entrepreneurs or business executives.[30]
- They typically invest in companies that operate in industries, or focus on technologies, with which they are personally familiar.[31]
- Angels are much more likely than venture capital firms to invest in a start-up's early stages.

In the United States, angels and venture capital firms generally have a complementary relationship, where the angels provide 'a kind of "farm system of venture portfolios."'[32] Angels provide the early-stage finance, and potentially the managerial experience (based on their prior business experience), to enable the start-ups to grow to a point where they may be attractive to venture capital firms. The result is that a healthy angel market is a valuable, and possibly necessary, part of well-functioning venture capital and start-up markets.[33] Without a well-functioning angel market, capital constraints will inappropriately limit the supply of start-ups that can eventually obtain venture capital financing.

China does not appear to have a robust angel finance market. We say 'appear' because angel markets are less formal, and the subject of much less study and attention, than professional venture capital markets. As a result, it is very difficult to obtain anything close to accurate information about a country's angel market, which is often referred to as the 'invisible' venture capital market.[34] We feel relatively confident, however, in assuming that China possesses a weak angel market. Typical preconditions for a vibrant angel market are:

- A large pool of former entrepreneurs or business executives who have successfully developed/managed companies and who have entered a period in their life when they have enough excess funds and time to devote to angel investing.
- Sufficient mechanisms exist to efficiently link potential angels with worthy start-ups.
- Effective practices develop for screening and managing investments in early-stage start-ups and become widely known among the pool of potential angel investors.
- Angels, entrepreneurs and lawyers develop experience in the best methods (e.g., should convertible preferred stock be used?) for structuring and documenting early-stage start-up investments.

Developing a vibrant angel finance market that can competently provide investment capital and valuable managerial experience to early-stage start-ups in China will be critical to the future success of China's start-up industry. Unfortunately, developing such a market is a complex undertaking that does not lend itself to quick fixes and easy solutions. Why does China not currently have a vibrant angel finance market? Is it simply a function of time? Will an angel finance market naturally develop as more Chinese entrepreneurs develop successful companies and eventually retire? Are there policies or regulations that are restricting the development of an angel finance market? Could universities help by offering more courses in entrepreneurialism that can excite future entrepreneurs about the possibility of acting as angels later in their careers? We do not have the answer to any of these questions. Right now, the most important step that can be taken to improve China's angel finance market is to highlight its importance and the crucial role it plays in the creation and development of start-ups in China. Identifying the problem will allow for policymakers, academics and other innovation system actors to thoughtfully develop solutions for a more vibrant angel finance market.

3.3 Continue to Experiment

We have focused on one issue (i.e., the funding gap) that dampers the creation and development of start-ups and one solution (i.e., the angel market) to that problem. We do not, however, want to leave the impression that the funding gap is the only potential impediment to China's start-up industry or that a more robust angel market is the only solution to the funding gap. There are a variety of problems that could harm China's start-up industry, including a number of potential regulatory problems. For example: is China's competition law being employed in a manner that allows start-ups to compete against larger companies? Are China's business entity laws and business licensing regulations sufficiently entrepreneur-friendly to encourage the creation of start-ups, or do such laws impose such heavy regulatory burdens that they discourage new company formation? Is China's tax regime properly supportive of entrepreneurial start-ups? In each case, these types of regulatory issues must constantly be monitored to ensure a proper environment for start-ups and they also require balancing a number of competing interests.

The important thing for Chinese policymakers to understand is that there are few, if any, absolutes when trying to develop an optimal environment for start-ups. China's experimental approach to developing its market-based innovation system (see Chapter 2) provides a useful model to follow as China continues to develop its start-up environment. Let us consider again the funding gap problem. Regulatory or tax incentives could be used in an experimental fashion to try and coax venture capital firms to conduct more early-stage start-up investments – e.g., venture capital firms that invest a certain amount of money (or a certain percentage of their investment portfolio) in early-stage companies could receive more favorable tax treatment. If the incentives provide the desired results without generating any inappropriate problems, the incentives can be continued. If not, then they are ended. One of the more admirable features of Chinese economic policymaking during the reform era has been the ability of policymakers to recognize (i) that they do not have perfect information when implementing a policy and (ii) the need to monitor the results of policymaking efforts and adapt when the desired results are not achieved. Such an approach will be optimal for developing an ideal start-up environment.

4. INCREASE AWARENESS OF THE ROLE THAT UNIVERSITY TECHNOLOGY COMMERCIALIZATION CAN PLAY IN CHINA'S ECONOMY

Our last recommendation is also our simplest. China should promote greater awareness of the valuable role that university technology commercialization can play in China's economy. Such awareness can help to incentivize the various actors in the university technology commercialization system to pursue valuable technology commercialization strategies, and to do so more efficiently. An effective university technology commercialization system requires a variety of different actors to each pursue individual strategies.

Understanding the benefits that can be generated by university technology commercialization, including the profit potential, can help improve and coordinate the decision making for each of these various actors. Various government agencies need to make numerous R&D funding decisions that require choosing amongst competing proposals. Greater awareness of technology commercialization can allow government funders to include commercialization potential as part of the funding criteria, which helps to better focus government R&D funds. University researchers need to choose between competing R&D opportunities. Understanding that commercialization opportunities are available can help incentivize university researchers to choose those R&D opportunities that have explicit commercial potential. Companies need to build technology relationships with universities. Ideally, some businesses will even begin to develop business strategies around acquiring university technology. Understanding the profit potential that can derive from such relationships/strategies can incentivize companies to pursue them. University TTOs need to market university-developed technology to the business sector. Greater awareness can reduce the costs involved with finding potential commercial partners. Finally, university administrators need to build institutions within the university to support the commercialization process (e.g., taking into account the commercialization potential of researchers when making hiring decisions) and create a technology commercialization culture. Obviously, those administrators need to be aware of, and understand, the process if they are to build effective institutions.

There are a number of common techniques that can be used to increase commercialization awareness. We would like to make a few suggestions:

- *Include university technology commercialization courses in science, business and law curriculums:* developing a truly robust university technology commercialization system is a long-term project. Just as China did not become a market-based economy overnight, neither will its universities commercialize technology at a level that will consistently drive economic growth overnight. It could take ten years, 20 years, or even more before China's universities truly begin to play that role. It is important, therefore, to take a long-term view of university technology commercialization. Seeds need to be planted today, however, to achieve that ultimate success.

One of the most valuable fields in which to plant such seeds is the minds of China's next generation of scientists, entrepreneurs and lawyers.

- *Promote university technology commercialization workshops for university administrators and government officials:* the key to educating university administrators and government officials is to help them to develop appropriate expectations about the benefits they can reasonably expect to achieve both in the long term and the short term. Understanding the long-term benefits helps to keep key decision makers motivated to keep pushing for university technology commercialization capacity to be developed. Understanding the limitations on the short-term benefits is equally important, as it allows decision makers to avoid costly and unrealistic over-investment in technology commercialization efforts. If too much money is lost in the short term due to unrealistic expectations, there is a real risk that key decision makers will incorrectly lose faith in the ultimate benefits from university technology commercialization and stop funding the efforts.

- *Universities should host networking events that bring together university researchers, high-technology businesses, potential investors, and IP lawyers:* universities can serve as convenient meeting points for many of the most important actors in an innovation system and thereby help to create productive relationships between them. We find that few universities take full advantage of this 'convenient meeting point' feature. Consider the following simple example: A research university hosts an event on an important technology topic. The university can employ a number of easy and inexpensive techniques to ensure attendance from a significant crowd of companies and potential investors. The university can have some of its best researchers talk about new developments in the field. The university can also ask some of the more important companies and investors to make presentations. In addition to getting a critical mass of people in the same room and hoping that fortuitous connections occur between such individuals, the university is well positioned to actually engineer these contacts. Information about the strengths and needs of each of the event participants can be collected before the event, and a few university employees can be charged with ensuring that individuals whose needs match with others' strengths actually meet. This is just one of many simple examples that a university can pursue to better link its local innovation environment.

- *An association of Chinese TTOs could play a primary role in increasing awareness:* we would expect that one of the roles of an association of Chinese TTOs would be to develop best practices training and education that could be broadly disseminated.

5. BOTTOM LINE

University-developed technology has the potential to make substantial contributions to China's economic development in the next few decades. All the basic ingredients are in place – or are in the process of developing (e.g., R&D capacity of Chinese companies) – for a meaningful, Bayh–Dole-style university technology commercialization system. China has developed a market-based economy and a market-based innovation system that facilitate university/industry technology exchange. The R&D capacity at China's universities – at least its elite universities – is improving, and their linkages with industry are developing. Questions persist about the R&D capacity of China's business sector and its ability to serve as an optimal technology transfer partner, but that problem should improve over time. China's legal system has also evolved to a point where it is more than sufficient to support efficient university technology commercialization. China's legal system supports market-based transactions, protects IP, and provides an incentive system that is comparable to the United States' Bayh–Dole approach to commercializing university-developed technology. Finally, market intermediaries such as venture capital firms are developing and beginning to take meaningful roles in the commercialization process.

With the broad foundation established, the challenges that lie ahead for university technology commercialization are likely to be much more of a specialized nature.

- Improving university IP management
- Finding ways to engage more universities in technology commercialization
- Continuing to increase the connections and productive exchanges between the various parties in the university technology commercialization system
- Developing missing intermediaries (e.g., angel investors) that can increase the efficiency of the market for university technology

If China simply continues in its current trajectory, we expect technology commercialization to grow in China's elite universities – although such growth may be slower than it could be. We would expect such growth to flow from growth trends in the R&D capacity of these elite universities, expected growth in business sector R&D capacity, continued growth in intermediaries, and greater experience by technology commercialization participants. By working on the details, however, China has the ability to not only significantly accelerate the growth of university technology commercialization, it also has the ability to spread its benefits to more Chinese universities and regions.

Conclusion: what does it mean for the rest of the world if China gets things right?

> I love being from the Third World because it represents such a marvelous challenge – that of making a transition to a market-based capitalist system that respects people's desires and beliefs. When capital is a success story not only in the West but everywhere, we can move beyond the limits of the physical world and use our minds to soar into the future.
>
> – Henry De Soto, in 'The Mystery of Capital'[1]

On the face of it, this book is about university technology commercialization in China. On a deeper level, however, this book has been about the transformation of a developing country to a point where it can begin to meaningfully engage with the knowledge-based world as an innovative country. China's progress in university technology commercialization is just one indication of China's ongoing transformation into an innovative country. Developing a robust university technology commercialization system could go a long way to satisfying China's desire to create meaningful domestic innovation capacity that can eventually provide the basis for China's future innovative growth.

Developing countries around the world want nothing more than to find a way to introduce meaningful technology-based economic development into their economies. For most of these countries, their universities and GRIs are likely to hold some of the country's strongest R&D capacity. For technology-based economic development to become a reality, therefore, universities and GRIs will almost certainly have to play a leading role in the effort. Finding the best way to unlock the technology commercialization capacity of developing-country universities and GRIs has so far proven to be elusive. It is not for lack of trying. Lots of ink has been spilt on the topic, workshops have been held, and international experts have been brought in to look at specific countries and their specific problems. Developing countries, as well as their universities and GRIs, are constantly looking for workable models to make their universities and GRIs more relevant for economic development.

One of the common threads underlying these efforts is that, in most cases, developing countries are seeking advice solely from the West and Western experts. It is unclear, however, whether the West has been very successful in providing this advice. Personally, we are concerned that the West has a near monopoly on economic and innovation development advice. Certainly the West provides a shining example of the benefits that can be derived from a market-based, knowledge economy. But, that does not necessarily mean the West always provides the best example of how to make the transition. In fact, this strikes us as being similar to the Silicon Valley/Route 128 paradox that we discussed in Chapter 9 (Section 2). A successful China could increase the diversity of opinions on how to conduct development, as China has not followed a purely Western approach.

We are concerned that Western experts and their models are too frequently based on Western 'best practices,' and such best practices fail to adequately appreciate the extent of the developing countries' innovation system idiosyncrasies and weaknesses. One of the more infamous examples of such failure by Western experts is economic reform of post-communist Russia that is commonly referred to as 'shock therapy.' The basic idea behind shock therapy was quite simple. Following an initial stabilization period, Russia's formerly planned economy was transformed into a market-based economy almost instantaneously, rather than in a more gradual fashion. Government controls over the economy (e.g., price controls, subsidies to state-owned enterprises, and employment controls) were quickly eliminated while state-owned assets were rapidly privatized.[2] The belief was that a properly-functioning market economy would naturally arise out of the newly liberalized environment.[3] Economists at the time understood that there would be a fair amount of pain involved in the transition, but akin to quickly ripping a bandage off a hairy arm, it was thought the preferred approach was to take that pain quickly and get on with the business of building a market-based economy. Shock therapy turned out to be very harmful to Russia's economy and its development. In their enthusiasm to create an ideal market environment, Western experts appear to have missed the fact that the various structures and intermediaries that support healthy markets in the West did not yet exist in Russia. Western experts saw the big picture – the need to move from a planned economy to a market-based economy – but they failed to appreciate that all of Russia's more subtle weaknesses meant markets would not quickly develop. Unfortunately, recognizing such subtle weaknesses could be the key to developing a successful innovation system.

On a similar note, we find that Western experts have a tendency to project their particular values onto developing countries and try to weave those values into their assessment of a county's problems. It is not uncommon, for example, for Western experts to express concern that China's approach to free speech or its emphasis on a stable society (e.g., 'hammer down the nail that sticks out') are antithetical to being an innovative nation because these policies stifle creativity.[4] How can China ask its researchers and entrepreneurs to engage in the pursuit of 'creative destruction,' while also communicating the need for stability and trust in authority? We believe this type of criticism has a lot of intuitive appeal for Westerners, because it is well-aligned with fundamental Western values. It is also possible – however – that China's focus on stability could be instrumental to its success so far in developing its innovation system. As far as we can tell, developing an innovation system is very much a long-term effort. It strikes us as being very possible that the value of stability in building such a long-term system may be undervalued by many Western experts.

All told, developing countries could benefit from a greater diversity of opinion on how to improve their innovation systems. If China is successful in developing an innovative nation that includes a robust university technology commercialization system, it will have made one of the most dramatic economic transformations in history. In 1978, China was one of the poorest countries in the world; it did not have a market-based economy, and its educational, R&D and legal institutions were all in disarray. It is hard to imagine a country having a more difficult starting point for creating an innovative nation. And yet, just over 30 years later, that is exactly what we are discussing with regard to China. For countries like Argentina, Brazil, India, Mexico, and South Africa – to name just a few – that are all working to create more innovation-based economies, China and its experts could prove to be a valuable source of information on how to make that transition. China has not developed into an innovative nation yet, nor has it developed a robust university technology commercialization system. But China is at a place where we can realistically talk about rooting for China to succeed. China's success will not only be important for the 1.3 billion people living in China, it could also prove to be the key for many in the developing world.

Notes

PREFACE

1. According to World Bank figures, China is the third largest economy in the world based on 2008 gross domestic product (nominal). World Development Indicators database, World Bank (July 1, 2009), available at http://siteresources.worldbank.org/DATASTATISTICS/Resources/GDP.pdf.
2. The State Council, The National Medium- and Long-Term Program for Science and Technology Development (2006–2020) – An Outline, *Section II.2, Guiding Principles, Development Goals, and General Deployment – Development Goals,* English version available at www.cstec.org/uploads/files/National%20Outline%20for%20Medium%20and%20Long%20Term%20S&T%20Development.doc.
3. John Mauldin, 'China: One Coin, Two Faces', *Thoughts From the Frontline,* Jan. 21, 2005, available at www.frontlinethoughts.com/article.asp?id=mwo012105.

CHAPTER 1

1. The World Bank (1998/99), *World Development Report 1998 – Knowledge for Development,* 16.
2. Paul M. Romer (2007), 'Economic Growth', *The Concise Encylopedia of Economics* (ed. David R. Henderson).
3. Ross Gittell & John Orcutt (2010), *New Hampshire in the Innovation Economy: A Plan to Increase Innovation and Technology-Based Economic Development in New Hampshire,* 17, available at www.epscor.unh.edu/.
4. *See* discussion *infra* Chapter 7.
5. Risaburo Nezu (2005), 'Technology Transfer, Intellectual Property and Effective University-Industry Partnerships: The Experience of China, India, Japan, Philippines, The Republic of Korea, Singapore and Thailand', *World Intellectual Property Organization Report,* 4–5.
6. Provisional Regulations of the State Council on Technology Transfer were promulgated on January 10, 1985. English language version available at http://www.novexcn.com/technology_transfer.html.
7. Sections 4 and 7 of the Provisional Technology Transfer Regulations.
8. *See* Robert Kneller (1999), 'Ownership Rights to University Inventions in Japan and China', *CASRIP Publication Series: Streamlining Int'l Intellectual Property,* pp. 160 and 162, available at http://www.law.washington.edu/Casrip/Symposium/Number5/pub5atcl18.pdf.
9. Charles Edquist, *The Systems of Innovation Approach and Innovation Policy: An Account of the State of the Art,* lead paper presented at the DRUID Conference, Aalborg (June 12–15, 2001), p. 2, available at www.druid.dk/conferences/nw/paper1/edquist.pdf.
10. *Id.* at 2–6.
11. Richard R. Nelson and Nathan Rosenberg (1993), 'Technical Innovation and National Systems', in Richard R. Nelson (ed.), *National Innovation Systems – A Comparative Analysis,* pp. 3–21.
12. *Id.* at 10. The business sector's dominant role should not be surprising, as it is the actor that is most directly able to capture the economic benefit from technological advances.

The business sector's close proximity to the general public's needs and desires is equally important. As the actor that most directly profits (or loses money) based on the technology decisions it pursues, the business sector receives extremely valuable information on what technologies are most desired by the public. This discipline of dealing with customers helps to incentivize the business sector to better focus its resources on the most valuable technology projects.

13. *See* Edquist, *supra* note 9, at 5. *See also* Eric Garduño (2004), 'South African University Technology Transfer: A Comparative Analysis', *International Intellectual Property Institute*, pp. 2–3.
14. *See* Edquist, *supra* note 9, at 2–3.
15. *See id.* at 3.
16. North was awarded the Nobel Memorial Prize in Economics in 1993 largely for this work on institutions and their role in economics. While not one of the five Nobel Prizes established by Alfred Nobel's will in 1895 (physics, chemistry, physiology or medicine, literature and for peace), the economics prize is commonly referred to as a Nobel prize. In 1968, the Bank of Sweden donated the funds to create the Sveriges Riksbank Prize in Economic Sciences in Memory of Alfred Nobel, and the economic prize is administered by the Royal Swedish Academy of Sciences just like the Nobel prizes in physics and chemistry. *Source:* the official website of the Nobel Foundation, http://nobelprize.org.
17. Douglass C. North, Lecture to the Memory of Alfred Nobel, December 9, 1993 (the lecture North gave when he received the Sveriges Riksbank Prize in Economic Sciences in Memory of Alfred Nobel), available at http://nobelprize.org/nobel_prizes/economics/laureates/1993/north-lecture.html. *See also* Douglass C. North (1990), *Institutions, Institutional Change and Economic Performance*, pp. 3–10.
18. Ian P. Bindloss, 'Contributions of Physics to the Information Age', *Dept. of Physics, UCLA*, available at http://www.physics.ucla.edu/~ianb/history/.
19. Pub. L. No. 96-517, 94 Stat. 3015, 3019–27 (1980) (codified at 35 U.S.C. §§ 200–211).
20. 'Innovation's Golden Goose', *The Economist*, December 12, 2002.
21. President John F. Kennedy issued a memorandum dated October 10, 1963 – 3 C.F.R. 861 (1959–1963) (Memorandum for the Heads of Executive Departments and Agencies) (the Kennedy Patent Policy) – that required the federal government in certain instances to provide exclusive licenses of federally funded inventions to the business sector. Sean M. O'Connor, *Historical Context of U.S. Bayh–Dole Act: Implications for Indian Government Funded Research Patent Policy*, available at SSRN: http://ssrn.com/abstract=1265343.
22. Federal Council for Science and Technology (1978), *Report on Government Patent Policy – Combined December 31, 1973, December 31, 1974, December 31, 1975, and September 30, 1976*, 403. *But see* David Mowery, Richard Nelson, Bhaven Sampat and Arvids Ziedonis (2004), *Ivory Tower and Industrial Innovation: University–Industry Technology Transfer Before and After the Bayh–Dole Act* 90–91 for a criticism of the validity of the 27 500/4.5% statistic (which is typically reported as 28 000/5%).
23. Garduño (2004), *supra* note 13 at 7.
24. Senator Birch Bayh, 'Bayh–Dole +25', *Innovation: America's Journal of Technology Commercialization* (April/May 2005), www.innovation-america.org/archive.php?articleID=89.
25. Donald Clarke, Peter Murrell & Susan Whiting (2008), 'The Role of Law in China's Economic Development', in Loren Brandt and Thomas G. Rawski (eds), *China's Great Economic Transformation*, p. 375.
26. *See* Gregg Graff (2007), 'Echoes of Bayh–Dole? A Survey of IP and Technology Transfer Policies in Emerging and Developing Economies', in Anatole Krattiger et al. (eds), *Intellectual Property Management in Health and Agricultural Innovation – A Handbook of Best Practices*, pp. 169, 170.
27. Garduño, *supra* note 13, at 17–18, provides a nice summary of the economic incentives that IP protection provides to inventors.

CHAPTER 2

1. Joe Studwell (2002), *The China Dream*, p.26.
2. Gregory C. Chow (2007), *China's Economic Transformation*, 2nd ed., p.29.
3. *Id.*
4. Robert A. Hillman (1997), *The Richness of Contract Law: an Analysis and Critique of Contemporary Theories of Contract Law*, p.214.
5. Donald Clarke, Peter Murrell & Susan Whiting (2008), 'The Role of Law in China's Economic Development', in Loren Brandt and Thomas G. Rawski (eds), *China's Great Economic Transformation*, p.375.
6. Xielin Liu and Steven White (2001), 'Comparing Innovation Systems: A Framework and Application to China's Transitional Context', 30 *Research Policy* 1091, 1097.
7. AnnaLee Saxenian and Xiaohong Quon (2005), 'Government and Guanxi: The Chinese Software Industry in Transition', in Simon Commander (ed.) *The Software Industry in Emerging Markets*, p.77.
8. Wei Hong (2005), *Technology Transfer of Chinese Universities: Forms and Implications*, Paper presented at the annual meeting of the American Sociological Association, 2 (Aug 12), available at www.allacademic.com/meta/p21558_index.html. It should be noted that in 1962, however, a few select universities were brought into China's national research system. At that time, the Education Ministry approved 18 research units in 11 universities and directed that their research should be aimed at applied, rather than basic, research.
9. Liu and White, *supra* note 6, at 1094, 1097. *But see* Ministry of Science and Technology at www.most.gov.cn/kjfz/kjlc/ (in Chinese), which states: 'China established an S&T system formed by five S&T research parts as follows: CAS, *universities*, research units under departments (ministries) of the State Council, local science research units and defense science research units' (emphasis added).
10. Liu and White, supra. note 6, at 1094 and 1097.
11. *See* id at 1097.
12. Xielin Liu and Nannan Lundin (2008), 'Toward a Market-Based Open Innovation System of China', in Reinhard Meckl (ed.), *Technology and Innovation Management: Theories, Methods and Practices from Germany and China*, p.18.
13. Liu and White, *supra* note 6, at 1098.
14. Even without the impact of a forced separation of research from industry, business leaders around the world struggle with how heavily to invest in R&D. R&D tends to be expensive, while its benefits are neither immediate nor certain (due to the risky nature of technological research). When faced with immediate and concrete R&D costs versus future and less certain benefits, it is not surprising that business leaders struggle with the 'how much to invest in R&D' decision. For Chinese business leaders, this R&D decision has been made even more difficult because they have not been operating in a business environment where R&D is central to the business culture. It is much easier to make an expensive, but uncertain, decision when one is operating in an environment where examples abound of the wisdom of such decision making. If anything, Chinese business leaders were exposed to the opposite influence, as their experience was of an R&D apparatus that was not properly in line with the needs of the business sector and its customers. None of this is meant to suggest that China's business leaders cannot, or will not, adapt to a more R&D-focused environment. It does help to explain, however, why it is taking so long for such a shift to take hold. Business leaders are being asked to take an expensive leap of faith without having developed full confidence in the wisdom of the strategy. Such confidence should develop as more Chinese companies benefit from the advantages of privately-funded R&D, but that is a process that takes time.
15. Saxenian & Quon, *supra* note 7, at 76–77.
16. *Id.* at 77.
17. Liu and White, *supra* note 6, at 1098.
18. Economist Joseph A. Schumpeter noted more than a half century ago that a healthy

economy is a dynamic organism that is constantly in a state of change and renewal. Joseph A. Schumpeter (1950), *Capitalism, Socialism, and Democracy*, p.83. Schumpeter described the process as one of 'Creative Destruction' whereby competition and innovation constantly revolutionize the economy from within – 'incessantly destroying the old one, incessantly creating a new one.' *Id.* By seeking innovations to render their competitors obsolete, entrepreneurs create new products, markets, processes for doing business, and even new industries, while old inefficient ones are destroyed. These newly created ventures must be more innovative and productive than their already established competitors in order to compete, which has the added benefit of forcing the established competitors to improve. Established competitors, as well as entire industries, that cannot meet the increased competition and innovations are forced out of business, which causes a constant renewal of the economy.

19. Liu and White, *supra* note 6, at 1098.
20. Id.
21. Daniel R. Fung (2008), 'The Rise of China: Political and Economic Implications', *Occasional Papers – Dean Rusk Center University of Georgia School of Law*, p.40, available at www.uga.edu/ruskcenter/pdfs/30th.pdf.
22. Saxenian & Quon, *supra* note 7, at 77.
23. *See also* William P. Alford (1995), *To Steal a Book is an Elegant Offense: Intellectual Property Law in Chinese Civilization*, p.64 ('the professional endeavors of virtually all scientists, writers, and other intellectuals were disrupted, and numbers of them were sent to the countryside, imprisoned, or subject to physical abuse').
24. Chow, *supra* note 2, at 2.
25. M.J. Greeven (2004), 'The Evolution of High-Technology in China after 1978: Towards Technological Entrepreneurship', *ERIM Report Series: Research in Management*, p.7, available at http://papers.ssrn.com/sol3/papers.cfm?abstract_id=636798.
26. International Development Research Centre (1997), *A Decade of Reform, Science and Technology Policy in China* [hereinafter IDRC Report], Chap. 7 – The High Technology Sector, available at www.idrc.ca/en/ev-55199-201-1-DO_TOPIC.html. Shulin Gu (1999), *China's Industrial Technology: Market Reform and Organizational Change*, p.13.
27. IDRC Report, *supra* note 26, at Chap. 7.
28. *See* Gu, *supra* note 26, at 14.
29. *Id.*, at 14.
30. *Id.*, at 12. IDRC Report, *supra* note 26, at Chap. 7.
31. *See* Embassy of the People's Republic of China in the Republic of Indonesia, *China's Science and Technology Development Summary*, available at http://id.china-embassy.org/eng/whjy/kjjl/t87396.htm.
32. *See* Gu, *supra* note 26, at 16.
33. The website for the Ministry of Science and Technology includes a basic description of the Key Technology R&D Program at www.most.gov.cn/eng/programmes1/200610/t20061009_36224.htm.
34. *Id. See also* Greeven, *supra* note 25, at 7.
35. Shulin Gu & Bengt-Åke Lundvall (2006), *China's Innovation System and the Move Toward Harmonious Growth and Endogenous Innovation*, *DRUID Working Paper No. 06-7*, p.18, available at www.druid.dk/wp/pdf_files/06-07.pdf. IDRC Report, *supra* note 26, Chap. 7.
36. *Id.*
37. *Id.*
38. *Id.*
39. *Id.*
40. Cong Cao (2004), 'Zhongguancun and China's High-Tech Parks in Transition', 44 *Asian Survey* 647, 652–653.
41. Wang Xiaomin (2000), *Zhongguancun Science Park: A SWOT Analysis*, Institute of Southeast Asian Studies, Visiting Researchers Series No. 10, p.3, available at http://iaps.cass.cn/UploadFile/200510621950903.pdf.

42. Merle Goldman (1994), *Sowing the Seeds of Democracy in China: Political Reform in the Deng Xiaoping Era*, p.314.
43. IDRC Report, *supra* note 26, Chap. 7. Cao, *supra* note 40, at 652–653.
44. *Id.*, at 653.
45. Quan Xiahong (2007),'China's Drive to Become a Technological Power', in Wang Gung Wu and John Wong (eds), *Interpreting China's Development*, p.114.
46. Gu, *supra* note 26, at17–28. Saxenian & Quon, *supra* note 7, at 78. Greeven, *supra* note 25, at 7.
47. Saxenian and Quon, *supra* note 7, at 78.
48. *Id.*
49. 'In a conference presenting scientific findings of universities in 1982, 32 universities presented 500 innovations, half of which resulted from collaborations among universities, research institutes, and industry.' Hong, *supra* note 8, at 2 (citing to the Chinese Education Ministry, *The 50-year Development of Science and Technology in Chinese University* (1999)). While no mention was made of the commercial value or usefulness of the 500 innovations, it is noteworthy that at least 32 universities were conducting commercial R&D in 1982, compared to 11 in 1962.
50. *See* Gu and Lundvall, *supra* note 35, at 17.
51. Gu, *supra* note 26, at 20
52. *Id.*, at 19.
53. *Id.*, at 20–21.
54. Gu and Lundvall, *supra* note 35, at 18.
55. Gu, *supra* note 26, at 19.
56. *See e.g.*, Linda Jakobson (2007), 'China Aims High in Science and Technology', in Linda Jakobson (ed.), *Innovation with Chinese Characteristics – High-Tech Research in China*, p.23.
57. *See* Gu, *supra* note 26, at 22–25.
58. *Id.*
59. *Id.*
60. English version of the State Council's announcement obtained from Gu, *supra* note 26, at 29, outer square brackets in the original.
61. Carl J. Dahlman and Jean-Eric Aubert (2001), *China and the Knowledge Economy – Seizing the 21st Century*, p.124.
62. Gu, *supra* note 26, at 29–30.
63. *Id.*
64. *Id.*, at 30.
65. *Id.*
66. *Id.*, at 68.
67. *Id.*, at 69.
68. *Id.*, at 71.
69. Greeven, *supra* note 25, at 9.
70. See, e.g., 2007 National Technology Market Statistics Annual Report (May 2007), available at www.chinatorch.gov.cn/yjbg/yjbg/200707/3811.html (in Chinese).
71. Cao, *supra* note 40, at 666.
72. *See id.*, at 649–650. *See generally*, Paul Krugman (1991), 'Increasing Returns and Economic Geography', 99 *Journal of Political Economy*, 483.
73. For a thoughtful analysis of the development of China's high-technology parks and the challenges they face, *see* Cao, *supra* note 40.
74. Saxenian & Quon, *supra* note 7, at 79.
75. *Id.*
76. *See id.*, at 79–80.

CHAPTER 3

1. The State Council, The National Medium- and Long-Term Program for Science and Technology Development (2006–2020) – An Outline, Section II.2, Guiding Principles, Development Goals, and General Deployment – Development Goals, English version available at www.cstec.org/uploads/files/National%20Outline%20for%20Medium%20 and%20Long%20Term%20S&T%20Development.doc.
2. *See* Martin Schaaper (2009), 'Measuring China's Innovation System: National Specificities and International Comparisons', *OECD Science, Technology and Industry Working Papers*, p.17, available at www.oecd.org/dataoecd/15/55/42003188.pdf.
3. Yao Li, John Whalley, Shunming Zhang and Xiliang Zhao (2009), 'The Transformation of China's Higher Education System and its Global Implications', *The University of Nottingham China Policy Institute, Discussion Paper 43*, p.4, available at www. nottingham.ac.uk/shared/shared_cpi/documents/discussion_papers/Discussion_43_ Education_Li_Whalley.pdf.
4. *See* Schaaper, *supra* note 2, at 23.
5. *See e.g.*, *id.*
6. *Id.* at 19.
7. Carl J. Dahlman & Jean-Eric Aubert (2001), *China and the Knowledge Economy – Seizing the 21st Century*, p.126.
8. 'China's Research Institutes to Operate for Profits', *English People's Daily*, Sept. 22, 2000, available at http://english.peopledaily.com.cn/english/200009/22/eng20000922_51103. html.
9. Dahlman & Aubert, *supra* note 7, at 126.
10. Schaaper, *supra* note 2, at 19.
11. *Id.* at 20.
12. National Center for Education Development Research of the Ministry of Education of the P.R. of China (2007), *OECD Thematic Review of Tertiary Education – Background Report for the P.R. of China* 14 [hereinafter OECD Review of China's Tertiary Education System]. Uwe Brandenburg and Jiani Zhu, Higher Education in China in Light of Massification and Demographic Change, Arbeitspapier Nr. 97, p.18.
13. *Id.*, at 7.
14. Li, Whalley, Zhang & Zhao, *supra* note 3, at 5.
15. Linda Jakobson (2007), 'China Aims High in Science and Technology', in Linda Jakobson (ed.), *Innovation with Chinese Characteristics – High-Tech Research in China*, p.8.
16. OECD (2008), *OECD Science, Technology and Industry Outlook*, pp.49–50 ('On average, 25% of the degrees awarded at universities in the OECD area were in science-related fields (engineering, manufacturing and construction, life sciences and agriculture, mathematics and computing).').
17. MOST (2008), *China S&T Statistics Databook*.
18. *Id.*
19. Barry Naughton (2007), *The Chinese Economy: Transitions and Growth*, p.362.
20. *Id.*
21. National Science Foundation – Division of Science Resources Statistics, Data Table: Degrees Awarded 1966–2006, by degree level – all fields and science and engineering, available at www.nsf.gov/statistics/nsf08321/content.cfm?pub_id=3785&id=2.
22. Li, Whalley, Zhang and Zhao, *supra* note 3, at 5.
23. Survey of Earned Doctorates, Doctorate Recipients from United States Universities – Selected Tables 2007, pp.1–2, available at www.norc.org/projects/survey+of+earned+ doctorates.htm.
24. Li, Whalley, Zhang and Zhao, supra. note 3, at 8.
25. *Id.*
26. *Id.*
27. *Id.*, at 8–9.
28. Brandenburg and Zhu, *supra* note 12, at 37–39.

29.   'Project 211' institutions in 2007:

1. Anhui University
2. Beijing Foreign Studies University
3. Beijing Forestry University
4. Beijing Institute of Technology
5. Beijing Jiaotong University
6. Beijing Normal University
7. Beijing University of Aeronautics and Astronautics
8. Beijing University of Chemical Technology
9. Beijing University of Chinese Medicine
10. Beijing University of Posts and Telecommunications
11. Beijing University of Technology
12. Central Conservatory of Music
13. Central South University
14. Central University of Finance and Economics
15. Chang'an University
16. China Agricultural University
17. China Pharmaceutical University
18. China University of Geosciences
19. China University of Mining and Technology
20. China University of Petroleum
21. China University of Political Science and Law
22. Chongqing University
23. Communication University of China
24. Dalian Maritime University
25. Dalian University of Technology
26. Donghua University
27. East China Normal University
28. East China University of Science and Technology
29. Fourth Military Medical University
30. Fudan University
31. Fuzhou University
32. Guangxi University
33. Guangzhou University of Traditional Chinese Medicine
34. Guizhou University
35. Harbin Engineering University
36. Harbin Institute of Technology
37. Hebei University of Technology
38. Hefei University of Technology
39. Hohai University
40. Huazhong Agricultural University
41. Huazhong Normal University
42. Huazhong University of Science and Technology
43. Hunan Normal University
44. Hunan University
45. Inner Mongolia University
46. Jiangnan University
47. Jilin University
48. Jinan University
49. Lanzh47.ou University
50. Liaoning University
51. Minzu University of China (formerly known as the Central University for Nationalities)
52. Nanchang University
53. Nanjing Agricultural University
54. Nanjing Normal University
55. Nanjing University
56. Nanjing University of Aeronautics and Astronautics
57. Nanjing University of Science and Technology
58. Nankai University
59. National University of Defense Technology
60. North China Electric Power University
61. Northeast Agricultural University
62. Northeast Forestry University
63. Northeast Normal University
64. Northeastern University
65. Northwest A&F University
66. Northwest University
67. Northwestern Polytechnical University
68. Ocean University of China
69. Peking Union Medical College
70. Peking University
71. Renmin University of China
72. Second Military Medical University
73. Shandong University
74. Shanghai International Studies University
75. Shanghai Jiao Tong University
76. Shanghai University
77. Shanghai University of Finance and Economics
78. Sichuan Agricultural University
79. Sichuan University
80. South China Normal University
81. South China University of Technology
82. Southeast University
83. Southwest Jiaotong University

84. Southwestern University of Finance and Economics
85. Sun Yat-sen University
86. Suzhou University
87. Taiyuan University of Technology
88. Tianjin Medical University
89. Tianjin University
90. Tongji University
91. Tsinghua University
92. University of Electronic Science and Technology of China
93. University of International Business and Economics
94. University of Science and Technology Beijing
95. University of Science and Technology of China
96. Wuhan University
97. Wuhan University of Technology
98. Xiamen University
99. Xi'an Jiaotong University
100. Xidian University
101. Xinjiang University
102. Xinjiang Medical University
103. Yanbian University
104. Yunnan University
105. Zhejiang University
106. Zhengzhou University
107. Zhongnan University of Economics and Law

Source: *China Education Online College Entrance Channel* (in Chinese), available at http://gaokao.eol.cn/gxmd_2920/20080401/t20080401_288694.shtml.

30. 'Over 10 Billion Yuan to be Invested in "211 Project"', People's Daily Online, Mar. 26, 2008 available at http://english.people.com.cn/90001/6381319.html.
31. *Id.*
32. *Id.*
33. Li Lixu (2004), 'China's Higher Education Reform 1998–2003: A Summary', 5 *Asia Pacific Education Review* 14, 17.
34. *Id.*
35. *Id.*
36. Brandenburg and Zhu, *supra* note 12, at 40.
37. 'Times Higher Education – QS World University Rankings 2009 – Top Universities', *QS Top Universities*, available at www.topuniversities.com/university-rankings/world-university-rankings/2009/results. The six schools are Tsinghua University (ranked 49th), Peking University (ranked 52nd), Fudan University (ranked 103rd), Shanghai Jiaotang University (ranked 153rd), University of Science and Technology of China (ranked 154th), and Nanjing University (ranked 168th). *Id.*
38. National Science Foundation , Division of Science Resources Statistics (2007), *Asia's Rising Science and Technology Strength, Comparative Indicators for Asia, the European Union, and the United States*, p.22 available at www.nsf.gov/statistics/nsf07319/ [hereinafter NSF Report].
39. James Wilsdon and James Keeley (2007), *China: The Next Science Superpower*, p.16, available at www.demos.co.uk/files/China_Final.pdf.
40. *Id.*
41. *Id.*
42. From 2005 to 2007, higher education sector R&D personnel increased 11.9% from 227 000 to 254 000. MOST, *China S&T Statistics Databooks* (2008 and 2006). MOST (2008), *China S&T Statistics Databooks* is available at www.sts.org.cn/sjk/kjtjdt/data2008/cstsm08.htm. MOST (1998–2007), *China S&T Statistics Databooks* are available at www.most.gov.cn/eng/statistics/2007/index.htm.
43. *See* MOST (2008), *China S&T Statistics Databooks*. From 2000 to 2007, the number of Chinese students who returned from studying abroad increased nearly 400% from roughly 9000 returning students in 2000 to 44 000 in 2007.

Chinese students studying abroad and returnees (2000–2007)

	2000	2001	2002	2003	2004	2005	2006	2007
				(in 10000 persons)				
Chinese students studying abroad	3.9	8.4	12.5	11.7	11.5	11.9	13.4	14.4
Returnees	0.9	1.2	1.8	2	2.5	3.5	4.2	4.4

Sources: MOST (2008), *China S&T Statistics Databooks* (for the 2002–2007 data) and MOST (2006), *China S&T* Statistics Databook (for the 2000 and 2001 data).

44. Wilsdon and Keeley, *supra* note 39, at 17.
45. National Science Board, Science and Engineering Indicators 2008, 5–48 (2008), available at www.nsf.gov/statistics/seind08/ [NSB Report].
46. Wilsdon and Keeley, *supra* note 39, at 17 (citing as source, 'Nine Problems Challenge the 'Innovative State', *People's Daily* (Jan. 9, 2006)').
47. NSF Report, *supra* note 38, at 29.
48. *Id.* 'Citations assigned to geographic location by institutional address(es) listed on article and calculated on fractional-count basis, i.e., for cited articles with collaborating institutions from multiple regions or countries/economies, each receives fractional credit on basis of proportion of its participating institutions.' *Id.*
49. OECD (2007), *OECD Reviews of Innovation Policy: China – Synthesis Report*, p.40 [hereinafter China – Synthesis Report]. *See also*, NSF Report, *supra* note 38, at 26. International research collaboration consists of coauthored articles having at least one institutional author that is outside China. *Id.*, 27.
50. X. Zhu et al. (2004), 'Highly Cited Research and the Evaluation of a Research University: A Case Study of Peking University', 1974–2003, 60 *Scientometrics* 237, 244. *See also*, Wilsdon and Keeley, *supra* note 39, at 18.
51. Schaaper, *supra* note 2, at 21.
52. MOST (2008 and 2006), *China S&T Statistics Databooks*.
53. MOST (2008 and 2004), *China S&T Statistics Databooks*.
54. NSB Report, *supra* note 45, at 4–43.
55. MOST (2008), *China S&T Statistics Databook*.
56. MOST (1999), *China S&T Statistics Databook*.
57. Jakobson, *supra* note 15, at 12.
58. Richard P. Suttmeier, Cong Cao and Denis F. Simon (2006), 'China's Innovation Challenge and the Remaking of the Chinese Academy of Sciences', *Innovations*, p.90, available at www.policyinnovations.org/ideas/policy_library/data/ChinasInnovation Challenge/_res/id=sa_File1/INNOV0103_p78-97_suttmeier.pdf.
59. Jakobson, *supra* note 15, at 12–14.
60. *Id.*, at 13–14. Jakobson's Nasdaq assertion is based on an interview she conducted with Professor Martin Kenney on March 29, 2006.
61. *Id.*, at 14.
62. Schaaper, *supra* note 2, at 28.
63. Jakobson, *supra* note 15, at 13.
64. Joint Research Centre's Institute for Prospective Technological Studies (2009), *2009 EU R&D Investment Scoreboard*, Dataset – R&D ranking of the top 1000 EU companies by country, available at http://iri.jrc.ec.europa.eu/research/scoreboard_2009.htm. The 15 Chinese companies are:

Chinese companies in the 2009 top 1000 non-EU R&D investors

Company	Rank	Sector	Company	Rank	Industry
PetroChina	76	Oil & Gas Producers	Semiconductor Manufacturing	530	Semiconductors
ZTE	133	Telecommunications Equipment	Tencent	627	Internet
China Petroleum and Chemical	143	Oil & Gas Producers	Dongfang Electric	708	Industrial Machinery
China Railway Construction	259	Construction & Materials	China Telecom	715	Fixed Line Communications
China Coal Energy	372	Mining	Harbin Power Equipment	735	Industrial Machinery
China Communications Construction	378	Construction & Materials	Weichai Power	851	Automobiles & Parts
BYD	402	Electronic Equipment	China Railway	964	Construction & Materials
China South Locomotive	403	Commercial Vehicles & Trucks			

65. Schaaper, *supra* note 2, at 4.
66. *See* Jakobson, *supra* note 15, at 13. *But see* Schaaper, *supra* note 2, at 44. Schaaper presents statistics showing that China's overall expenditures on technology importation has been roughly $30 billion per year from 1995–2005, while overall R&D expenditures have steadily increased – in particular since 1999. Schaaper explains that such statistics imply 'that a complimentary, rather than a substitute relationship between R&D expenditure and foreign technology imports seems to prevail at the current stage of S&T activities in China.'
67. Jakobson, *supra* note 15, at 15.
68. To be clear, a patent does not give its holder a right to 'practice' its invention. A patent only gives the holder the right to 'exclude' others from doing so.
69. *See* Section 1.1 of Chapter 5.
70. Article 2 of China's Patent Law.
71. Article 45 of China's Patent Law.
72. Wei Hong (2005), *Technology Transfer of Chinese Universities: Forms and Implications*, paper presented at the annual meeting of the American Sociological Association, p.5 (Aug 12), available at www.allacademic.com/meta/p21558_index.html.
73. Article 9 of China's Patent Law.
74. WIPO (2008), *World Patent Report – A Statistical Review*, 17, available at www.wipo.int/ipstats/en/statistics/patents/wipo_pub_931.html.
75. OECD, *Compendium of Patent Statistics – 2008*, p.6, available at www.oecd.org/datao.ecd/5/19/37569377.pdf.
76. OECD (1997), *National Innovation Systems*, p.9.
77. Schaaper, *supra* note 20.
78. *Id.*, 28.
79. Schaaper, *supra* note at 2, 28.
80. OECD (2008), *OECD Reviews of Innovation Policy: China*, pp.193–194.
81. *Id*, at 194.
82. OECD, *supra* note 80, at 194.

CHAPTER 4

1. Derek Bodde and Clarence Morris (1967), *Law in Imperial China – Exemplified by 190 Ching Dynasty Cases*, p.25 (translated from the Hsing-an hui-lan).
2. Peter Murrell (2003), 'The Relative Levels and the Character of Institutional Development in Transition Economies', in Nauro F. Campos and Jan Fidrmuc (eds), *Political Economy of Transition and Development: Institutions, Politics and Policies* (ZEI Studies in European Economics and Law), pp.50–58.
3. *See e.g.*, *World Development Report 2002, Building Institutions for Markets* (2002) [hereinafter *Building Institutions for Markets*]; Douglas W. Arner (2007), *Financial Stability, Economic Growth, and the Role of Law*, pp.13–22; Mwangi S. Kimenyi & John Mukum Mbaku (2003), 'Institutions and Economic Growth', in Mwangi S. Kimenyi, John Mukum Mbaku and Ngure Mwaniki (eds), *Restarting and Sustaining Economic Growth and Development in Africa*, pp.13–34.
4. *See* Stanley Lubman (2006), 'Looking for Law in China', 20 *Columbia Journal of Asian Law* 1, 4–5.
5. *See id.*, at 32 ('China's law schools, already politicized by the late 1950s, were closed at the beginning of the Cultural Revolution and were the last educational institutions to reopen, in 1979.'). In fact, a few law schools had reopened prior to 1979. Peking University's Law Department, for example, reopened in 1972 and had already begun a period of improvement in 1977 following the reform of the university admission examination system. *History of PKULS*, Peking University Law School website at http://en.law.pku.edu.cn/article_one.asp?MID=20091124146848&MenuId=20091124 132733&menuname=About.
6. OECD (2003), *Investment Policy Reviews: China – Progress and Reform Challenges* p.111 [hereinafter *China – Progress and Reform Challenges*].
7. Donald Clarke, Peter Murrell & Susan Whiting (2008), 'The Role of Law in China's Economic Development', in Loren Brandt and Thomas G. Rawski (eds), *China's Great Economic Transformation*, p.376.
8. *Id.*
9. *See id.*
10. *See* 'Communique of the Third Plenary Session of the 11th Central Committee of the Communist Party of China', *Beijing Review No. 52* (1978), available at www.bjreview. com.cn/nation/txt/2009-05/26/content_197538.htm.
11. 'The Four Modernizations, 1979–82', *China: A Country Study* (eds. Robert L. Worden, Andrea Matles Savada and Ronald E. Dolan) (1987), available at http:// countrystudies.us/china/.
12. *Id.*
13. *Id.*
14. Barry Naughton (2007), *The Chinese Economy – Transitions and Growth*, p.88.
15. *Id.*
16. Donald C. Clarke (2008), 'Legislating for a Market Economy in China', in Donald C. Clarke (ed.), *China's Legal System: New Developments, New Challenges*, p.15.
17. Clarke, Murrell & Whiting, *supra* note 7, at 389.
18. Naughton, *supra* note 14, at 97.
19. *Id.*
20. *Id.*
21. *Id.*
22. *Id.*
23. *Id.*
24. *See* Clarke, Murrell & Whiting, *supra* note 7, at 379.
25. Hong Shen, who attended Peking University's Law Department from 1979 to 1983 would like to thank her professors at Peking University for their tireless and skilled contributions to the development of China's legal system during the early phases of the reform era. During those early years, valuable assistance by those professors in

the law making process helped to improve the quality of the resulting laws and help the government to understand both the need for, and the value of, developing a more law-dependent economy. Their contributions are too frequently forgotten when considering China's early reform movement.

26.	Clarke, Murrell and Whiting, supra note 7 at 377.
27.	*See* Lubman, *supra* note 4, at 7.
28.	Clarke, Murrell & Whiting, *supra* note 7, at 377.
29.	*Id.*, at 378.
30.	Lubman, *supra* note 4, at 7; Clarke, Murrell & Whiting, *supra* note 7, at 377.
31.	Ge Hu (1986), 'The Implications of Enforcing the Bankruptcy System', *Guangming Daily*, p.3 (July 26) (cited by Clarke, Murrell & Whiting, *supra* note 7, at 377).
32.	Clarke, Murrell & Whiting, *supra* note 7, at 378.
33.	Article 2 of the Economic Contract Law (1981) ('Economic contracts are agreement between legal persons for the purpose of realizing certain economic goals and clarifying each other's rights and obligations.').
34.	'Although the ECL [Economic Contract Law] itself did not provide a definition of 'legal person,' the definition associated with the ECL was that a legal person must have an independent budget, independent cost accounting, and the right to possess capital. The legal person, then, is one that not only has the authority to commit finances to a contract but also has the capacity to be held financially accountable in the event of nonfulfillment.' Pitman B. Potter (1992), *The Economic Contract Law of China: Legitimation and Contract Autonomy in the PRC*, pp.31–32.
35.	*See* Clarke, Murrell & Whiting, *supra* note 7, at 389.
36.	Lubman, *supra* note 4, at 7.
37.	Foreign-Related Economic Contract Law of the PRC (1985).
38.	General Principles of Civil Law (GPCL), Article 1. An English version of the GPCL is available at www.fdi.gov.cn/pub/FDI_EN/Laws/law_en_info.jsp?docid=50982.
39.	Clarke, Murrell & Whiting, *supra* note 7, at 389.
40.	GPCL, Chapter II – Sections 1, 4 and 5, and Chapter III.
41.	GPCL, Chapter IV.
42.	Lubman, *supra* note 4, at 8.
43.	China's 1954 Constitutions provided two methods for owning the 'means of production.' Specifically, Article 5 of the 1954 Constitution stated:

> 'The ownership of the means of production is as follows: state ownership, which means ownership by the people as a whole; cooperative ownership, which means collective ownership of the working people; individual worker ownership; and capitalist ownership.'

Article 5 of China's 1975 Constitution refined this concept of ownership:

> 'There are currently two forms of ownership of the means of production in the People's Republic of China: the socialist ownership by the whole people and the socialist collective ownership by the working people.'

Article 5 of China's 1978 Constitution was identical to the version in the 1975 Constitution.
44.	1982 Constitution, Articles 6 and 7.
45.	1982 Constitution, Articles 8–10.
46.	1982 Constitution, Article 11.
47.	*Id.*
48.	Clarke, Murrell & Whiting, *supra* note 7, at 388.
49.	*Id.*, at 388–389.
50.	GPCL, Chapter III.
51.	Law on Chinese–Foreign-Equity Joint Ventures (1979).

52. Lubman, *supra* note 4, at 8.
53. Justin Yifu Lin (2005), 'Viability, Economic Transition and Reflection on NeoClassical Economics', 58 *Kyklos* 239, 240.
54. Clarke, Murrell & Whiting, *supra* note 7, at 404.
55. *Id.*
56. Naughton, *supra* note 14, at 91–92.
57. *Id.*, at 92.
58. *Id.*
59. *Id.*, at 99.
60. *Id.*, at 100.
61. Clarke, *supra* note 16, at 17.
62. *Id.*
63. Naughton, *supra* note 14, at 91 and 102–103.
64. *See generally* Clarke, *supra* note 16, at 17–29; and Naughton, *supra* note 14, at 102–103.
65. Articles 1–4 of the Property Law (2007).
66. Article 5 of the 1999 Constitution. Article 5 remained unchanged in the 2004 Constitution.
67. Gregory C. Chow (2007), *China's Economic Transformation*, 2nd ed, p.370.
68. *China – Progress and Reform Challenges*, *supra* note 6, at 111.
69. Information Office of the State Council (2008), *White Paper: China's Efforts and Achievements in Promoting the Rule of Law*, Chap. 6, available at http://english.gov.cn/2008-02/28/content_904901.htm.
70. *Id.*, at Chap. 7.
71. *Id.*
72. *See* Clarke, Murrell & Whiting, *supra* note 7, at 379.
73. For a nice statement of this concept, *see* Dan Harris (2007), 'China Needs More Lawyers. Hallelujah. Hallelujah', *China Law Blog* (Oct. 30), available at www.china-lawblog.com/2007/10/china_needs_more_lawyers_halle.html.
74. Albert H.Y. Chen (2009), *Legal Thought and Legal Development in the People's Republic of China 1949–2008* (March 28), available at SSRN: http://ssrn.com/abstract=1369782.
75. Yuwen Li (2002), 'Court Reform in China: Problems, Progress and Prospects', in Jianfu Chen, Yuwen Li and Jan Michiel Otto (eds) *Implementation of Law in the People's Republic of China* p.69.
76. Article 1 of the Law on Judges (English translation from Asian Legal Information Institute, Laws of the People's Republic of China, Judges Law, available at www.asianlii.org/cn/legis/cen/laws/jl93/).
77. Article 9 of the Law on Judges.
78. Article 12 of the Law on Judges.
79. Ji Weidong, *Judicial Reform in China and its Political Implications*, Paper presented at the annual meeting of the Research Committee of Sociology of Law, Paris (2005), 2, available at www.reds.msh-paris.fr/communication/docs/weidong.pdf.
80. *China – Progress and Reform Challenges*, *supra* note 6, at 112.
81. In John Orcutt's home state of New Hampshire, a judge resigned in 2008 after being suspended for three years by the New Hampshire Supreme Court for fraudulent and deceptive behavior. The judge, Rockingham County Superior Judge Patricia Coffey, was suspended for three years after the New Hampshire Supreme Court concluded 'she was complicit in a fraudulent property transfer' and was evasive and misleading to investigators. Norma Love, *Judge Coffey Resigns*, Seacoastonline.com (Apr. 21, 2008), available at www.seacoastonline.com/apps/pbcs.dll/article?AID=/20080421/NEWS/80421015 [hereinafter *Judge Coffey Resigns*]. Two years earlier, Judge Coffey was investigated by New Hampshire's judicial conduct committee for possibly sleeping during a trial 'that sent a man to jail for at least 23 years.' Josh Rogers (2006), 'Allegation of Sleeping Judge Raises Questions', *New Hampshire Public Radio* (Jan.

11), available at www.nhpr.org/node/10177. The panel investigating Judge Coffey ultimately concluded that she did not sleep, but 'briefly nodded off.' *Judge Coffey Resigns.*

82.   Clarke, Murrell & Whiting, *supra* note 7, at 399–400.
83.   For a detailed discussion of the debates surrounding the enactment of the Patent Law
       (1984), *see* Peter Ganea (2005), 'Patents, Utility Models and Designs', in Peter Ganea,
       Thomas Pattloch and Christopher Heath (eds), *Intellectual Property Law in China*,
       pp.1–8.
84.   Id., 4–8.
85.   *See* Chapter 5, section 1.
86.   Mo Zhang (2008), 'From Public to Private, The Newly Enacted Chinese Property Law
       and the Protection of Property Rights in China', 5 *Berkeley Bus. L.J.* 317.
87.   *See* Clarke, Murrell & Whiting, *supra* note 7, at 399. *See also* Zhang, *supra* note 86, at
       363.
88.   Clark, Murrell & Whiting, *supra* note 86, at 399–400.
89.   *See e.g.*, Zhang, *supra* note 7, at 357–363.
90.   Chapter 2 of the Contract Law (1999).
91.   Chapter 3 of the Contract Law (1999).
92.   Chapters 4 and 7 of the Contract Law (1999).
93.   Chapter 7 of the Contract Law (1999).
94.   Yadong Luo (2000), *Guanxi and Business*, p.2.
95.   Udo C. Braendle, Tanja Gasser & Juergen Noll (2005), *Corporate Governance in China
       – Is Economic Growth Potential Hindered by Guanxi?* 4–5, available at http://papers.
       ssrn.com/sol3/papers.cfm?abstract_id=710203.
96.   For a detailed discussion of the social networks and their role in economic activities,
       see Avner Greif (2006), *Institutions and the Path to the Modern Economy – Lessons
       from Medieval Trade.*
97.   *See also Building Institutions for Markets*, *supra* note 3, at 6.
98.   *Id.*
99.   For example, a 2005 study published by Susan Whiting, a political science professor at
       the University of Washington, found Chinese companies used written contracts with
       customers in 98.6% of the cases she reviewed and in 90.5% of cases with suppliers.
       Clarke, Murrell & Whiting, *supra* note 7, at 408 (describing Whiting's study).
100.   Dan Harris (2009), 'Topless Women, Rule of Law, and Perceptions of China', *China
       Law Blog* (Sept. 13), available at www.chinalawblog.com/2009/09/topless_women_
       rule_of_law_and.html.
101.   Randall Peerenboom (2002), *China's Long March toward Rule of Law*, p.1.
102.   Randall Peerenboom, *China's Legal System: A Bum Rap?*, UCLA International
       Institute, www.international.ucla.edu/article.asp?parentid=2878 [hereinafter Peeren-
       boom – *A Bum Rap?*].
103.   *Id.*
104.   Dan Harris (2009), 'China's First Foreign Nail House. Dude, Where's your Contract?'
       China Law Blog (Sept. 7), available at www.chinalawblog.com/2009/09/chinas_first_
       foreign_nail_hous.html.
105.   Peerenboom – *A Bum Rap?*, *supra* note 102.

CHAPTER 5

1.   Premier Wen Jiabao in a 2008 speech addressing a meeting of the Communist Party
      of China on the study and practice of scientific progress on development. *China's
      Intellectual Property Protection in 2008*, SIPO Website, available at www.sipo.gov.
      cn/sipo_English/laws/whitepapers/200904/t20090427_457167.html. We have slightly
      cleaned up the English translation of Premier Wen Jiabao's statement posted on the
      SIPO website.
2.   Eric Garduño (2004), *South African University Technology Transfer: A Comparative*

Analysis, International Intellectual Property Institute, pp.17–18, provides a nice summary of the economic incentives that IP protection provides to inventors.

3. *See* Paul M. Romer (1990), 'Endogenous Technological Change', 98 *Journal of Political Economy* S71, S74.
4. Douglass North (1981), *Structure and Change in Economic History*, p.164.
5. OECD (2003), *Investment Policy Reviews: China – Progress and Reform Challenges*, p.118 (2003) [hereinafter *China – Progress and Reform Challenges*].
6. Hua Guo (2007), 'IP Management at Chinese Universities', in Anatole Kratiger et al. (eds), *Intellectual Property Management in Health and Agricultural Innovation – A Handbook of Best Practices*, p.1674.
7. The 2008 amendment to the Patent Law went into effect on October 1, 2009.
8. The State Intellectual Property Office of the PRC (SIPO) provides an English version of the Patent Law at www.sipo.gov.cn/sipo_English/laws/lawsregulations/200804/t20080416_380327.html; and the Implementing Regulations at www.sipo.gov.cn/sipo_English/laws/lawsregulations/200804/t20080416_380326.html.
9. Article 2 of the Patent Law.
10. Rule 2 of the Implementing Regulations of the Patent Law.
11. Article 22 of the Patent Law. Article 22 provides the following definitions of novelty, inventiveness and practical applicability:

> 'Novelty means that the invention or utility model is not an existing technology, and prior to the date of application, no entity or individual has filed an application heretofore with the patent administrative department of the State Council for the identical invention or utility model and recorded it in the patent application documents or patent documents released after the said date of application.
> Inventiveness means that, as compared with the technology existing before the date of application, the invention has prominent substantive features and represents a notable progress and that the utility model has substantive features and represents progress.
> Practical applicability means that the invention or utility model can be made or used and can produce effective results.
> The term 'existing technology' as mentioned in this Law refers to the technologies known to the general public both at home and abroad prior to the date of application.'

12. Article 42 of the Patent Law.
13. Rule 2 of the Implementing Regulations of the Patent Law.
14. Article 22 of the Patent Law.
15. Article 42 of the Patent Law.
16. Rule 2 of the Implementing Regulations of the Patent Law.
17. Article 23 of the Patent Law. Article 23 describes the requirements of a design patent as follows:

> 'Any design for which a patent is granted shall not be attributed to the existing design, and no entity or individual has, before the date of application, filed an application with the patent administrative department of the State Council on the identical design and recorded it in the patent documents published after the date of application. As compared with the existing design or combination of the existing design features, the design for which a patent is granted shall have distinctive features. The patented design may not conflict with the lawful rights that have been obtained by any other person prior to the date of application. The term 'existing design' as used in this Law refers to a design known to the general public both at home and abroad prior to the date of application.'

18. Article 42 of the Patent Law.

19. Article 11 of the Patent Law.
20. *Id.*
21. Article 25(1) of the Patent Law.
22. Chapter 1, Section 4.1 of China's Guidelines for Patent Examination (the Guidelines).
23. *Id.*
24. Article 25(2) of the Patent Law.
25. Chapter 1, Section 4.2 of the Guidelines.
26. *Id.*
27. Wenping Chen & Xun Feng (2003), 'The China IP Focus 2003: How to Distinguish Patentable Subject Matter', *Managing Intellectual Property* (January), available at http://managingip.com/Article.aspx?ArticleID=473279.
28. Article 25(3)–(5) of the Patent Law.
29. Article 5 of the Patent Law.
30. Article 28 of the Patent Law.
31. Articles 29 and 30 of China's Patent Law. *But see* Article 20 of China's Patent Law which requires that Chinese parties that develop an invention/creation in China must report the invention to the Chinese patent office prior to filing a patent application in a foreign country.
32. Article 39 of the Patent Law.
33. Article 40 of the Patent Law.
34. Article 6 of the Patent Law and Rule 11 of the Implementing Regulations.
35. Article 16 of the Patent Law.
36. 'State Council OKs New Patent Law Implementing Regulations', *China IP News* (Feb. 1, 2010), available at www.chinaipr.gov.cn/news/headlines/607409.shtml.
37. Article 6 of the Patent Law.
38. Robert Kneller (1999), 'Ownership Rights to University Inventions in Japan and China', *CASRIP Publication Series: Streamlining Int'l Intellectual Property*, p.162, available at www.law.washington.edu/Casrip/Symposium/Number5/pub5atcl18.pdf.
39. Article 6 of the Patent Law.
40. Article 7 of the Copyright Law.
41. Articles 11 and 16 of the Copyright Law.
42. Article 16 of the Copyright Law.
43. *Id.*
44. *Id.*
45. *Id.*
46. Article 58 of the Copyright Law.
47. Article 13 of the Software Regulations.
48. *See e.g.,* United States Trade Representative (2008), *2008 Report to Congress on China's WTO Compliance*, pp.4–5 (2008), available at www.ustr.gov/sites/default/files/asset_upload_file192_15258.pdf; *China – Progress and Reform Challenges, supra* note 5, at 126–127; Joseph A. Massey (2006), 'The Emperor is Far Away: China's Enforcement of Intellectual Property Rights Protection', 7 *Chicago Journal of International Law* 231, 231; Justin Hughes (2006), *IP Enforcement in China, a Potential WTO Case, and U.S.–China Relations*, Written Statement before the U.S.–China Economic and Security Review Commission (June 8), p.2, available at www.uscc.gov/hearings/2006hearings/written_testimonies/06_06_08wrts/06_06_7_8_hughes_justin.pdf.
49. *See e.g.,* Peter K. Yu (2006), 'From Pirates to Partners (Part Two): Protecting Intellectual Property in Post-WTO China', 55 *Am. U. L. Rev.* 901, 927 (explaining that piracy numbers supplied by a self-interested trade group like the Business Software Alliance were unlikely to be taken at face value by the WTO Dispute Settlement Panel) [hereinafter Yu – *Pirates*]; and William P. Alford (1999), *To Steal a Book is an Elegant Offense: Intellectual Property Law in Chinese Civilization*, p.129 n. 13.

'The U.S. International Trade Commission . . . estimates that foreign infringement of American intellectual property costs this country more than 133 000 jobs and from U.S. $23.8 to U.S. $61 billion in lost profits annually, and that [China] and

[Taiwan] account for a sizable portion of that infringement. . . . These figures should not be taken at face value, as they are based on data supplied by domestic industries seeking government assistance against infringers and typically calculate losses by multiplying estimated instances of infringement by full list prices. Even assuming the accuracy of the estimates of the numbers of infringers, there is no reason to presume that each infringer would prefer to pay a list price rather than cease using the item in question, were these the only two alternatives available.' *Id.*

50. The English-language version of the Business Software Alliance website is located at www.bsa.org/country.aspx?sc_lang=en.
51. *See* Yu – *Pirates, supra* note 49, at 927, who provides the following explanation about the Second Annual Business Software Alliance Piracy Study (2005):

> '[E]verything is relative. If the study by the Business Software Alliance was accurate – that is a very big if – we should not ignore the fact that the United States has a software piracy rate of twenty-one percent while other developed countries, like France, Italy, and Spain, have piracy rates that range from the mid-forties to the low-fifties. A piracy rate of ninety for a country that did not have intellectual property laws twenty-five years ago is not as problematic as a rate of forty to fifty percent for a country that has had a well-established intellectual property system for more than two centuries'.

52. Peter Yu (2007), 'Intellectual Property, Economic Development, and the China Puzzle', in Daniel J. Gervais (ed.), *Intellectual Property, Trade and Development Strategies to Optimize Economic Development in a TRIPs Plus Era*, p.203 [hereinafter Yu – *China Puzzle*].
53. 'China in Transition: Regional Disparities have gone Beyond Acceptable Limits – The Path to an All-Around Well-Off Society Remains Distant', *Research Insititute of Economy, Trade & Industry, IAA* (2005), available at www.rieti.go.jp/en/china/05112901. html.
54. *See e.g.*, China's Eleventh Five-Year Plan, which includes 'promoting balanced development between urban and rural areas and among regions' as one of its guiding principles. National Development and Reform Commission of the People's Republic of China, *The Outline of the Eleventh Five Year Plan – For National Economic & Social Development of the People's Republic of China*, 4 and 20–23, available at http://en.ndrc. gov.cn/hot/t20060529_71334.htm.
55. Hu Angang (2003), 'How Will China Build a Well-Off Society for All of Its Citizens', *The China Business Review* (Mar.–Apr), pp.50–55.
56. *Id.*
57. Commission on Intellectual Property Rights (2003), *Integrating Intellectual Property Rights and Development Policy: Report of the Commission on Intellectual Property Rights*, p.147, available at www.iprcommission.org/papers/pdfs/final_report/ CIPRfullfinal.pdf.
58. Yu – *China Puzzle, supra* note 52, at 206.
59. *WTO Dispute Settlement, Dispute DS362, China — Measures Affecting the Protection and Enforcement of Intellectual Property Rights*, World Trade Organization (request for consultation filed by the United States on Apr. 10, 2007), available at www.wto.org/ english/tratop_e/dispu_e/cases_e/ds362_e.htm.
60. John Lee & Eion Murdock (2009), 'IP Enforcement in China', *Supplement – China IP Focus 2009* (7th ed.), available at www.managingip.com/Article/2176060/IP-enforcement-in-China.html.
61. *Id.*
62. *See id.*
63. *See id.*
64. Chapter VII of the Patent Law.

65. Lee & Murdock, *supra* note 60.
66. Lee & Murdock, *supra* note 60.
67. Lee & Murdock, *supra* note 60.
68. Tony Chen (2008), 'Beijing High Court Upholds Viagra Patent in China', *IP Perspectives*, p.30, available at www.jonesday.com/files/Publication/288b184e-c6ee-44 b5-800f-30838f34da54/Presentation/PublicationAttachment/aa464b25-7839-4af9-be34- 30d62faf4d56/Beijing_High_Court.pdf. Yu – Pirates, *supra* note 49, at 984. Jeffrey A. Andrews (2006), Pfizer's Viagra Patent and the Promise of Patent Protection in China, 28 *Loy. L.A. Int'l & Comp. L. Rev.* 1. Geoffrey K. Cooper (2004), *Patent Invalidation in Post-WTO China: Pfizer's Sildenafil Use*, JurisNotes.com 1, available at www.juris- notes.com/IP/articles/patentinvalidation.htm. Elizabeth Chien-Hale (2007), *Intellectual Property Aspects of Doing Business in China*, 1626 PLI/Corp, pp.109, 203.
69. Chen, *supra* note 68, at 30.
70. *Id.*
71. *Id.*
72. See *id.* See also WIPO Website, *Contracting Parties – WIPO Convention, Patent Cooperation Treaty* (which explains that China's entry into the PCT took force on Jan. 1, 1994), available at www.wipo.int/treaties/en/ShowResults.jsp?country_id=ALL&start_ year=ANY&end_year=ANY&search_what=C&treaty_id=1&treaty_id=6.
73. Chen, *supra* note 68, at 31.
74. *Id.*
75. *Id.* Andrews, *supra* note 68, at 30–31.
76. Jill Newton (2006), 'Viagra® Patent Appeal Case Clarifies Best Method Disclosure Requirements in Australia', *The Watermark Journal* 23, 1 (describing the find- ings of the full Federal Court in Pfizer Overseas Pharmaceuticals v Eli Lilly and Company [2005] FCA 67), available at www.iam-magazine.com/reports/Detail. aspx?g=0ad0f455-e891-469c-ba3b-71c9e50afdc0.
77. Andrews, *supra* note 68, at 21.
78. Chen, *supra* note 68, at 31.
79. Article 45 of China's Patent Law:

> 'Where, as of the announcement of the granting of the patent by the patent admin- istrative department of the State Council, any entity or individual considers that the granting of the said patent does not conform to the relevant provisions of the Patent Law, it or he may request the Patent Reexamination Board to invalidate the patent right.'

80. Articles 45 and 46 of the Patent Law.
81. Chen, *supra* note 68, at 31.
82. Yu – Pirates, *supra* note 49, at 984–985. Andrews, *supra* note 68, at 13–14.
83. Article 26 of the Patent Law.
84. Paul Mooney (2004), 'China Challenging Drug Patents', *The Scientist* (Aug. 20), avail- able at www.biomedcentral.com/news.20040820/02.
85. Article 27(1) of the Agreement on Trade-Related Aspects of Intellectual Property Rights (commonly referred to at the TRIPs Agreement), Apr. 15, 1994.
86. Naotaka Matsukata (2004), 'China's Counterfeit Commitment to Patents', *Fin. Times* (Aug. 5), p.17.
87. See e.g., Andrews, *supra* note 68, at 2.
88. Yu – *China Puzzle*, *supra* note 52, at 986.
89. Article 46 of the Patent Law provides in part:

> 'Where any party is dissatisfied with the decision of the Patent Reexamination Board on declaring a patent invalid or maintaining a patent, such party may, within three months of receipt of the notification, bring a lawsuit to the people's court. The people's court shall notify the opposite party in the procedures for requesting invalidation that it or he should participate in the litigation as a third party.'

90. Chen, *supra* note 68, at 31.
91. *Id.*, at 31–32.
92. *Id.*, at 32.
93. *Id.*
94. *Id.*
95. *Id.*
96. *Id.*

CHAPTER 6

1. OECD (2003), *Turning Science into Business: Patenting and Licensing at Public Research Organizations*, p.9.
2. Hua Guo (2007), 'IP Management at Chinese Universities', in Anatole Kratiger et al. (eds), *Intellectual Property Management in Health and Agricultural Innovation – a Handbook of Best Practices*, p.1675.
3. Promulgated on January 10, 1985. English language version available at www.novexcn.com/technology_transfer.html.
4. Sections 4 and 7 of the Provisional Technology Transfer Regulations.
5. Section 4 of the Provisional Technology Transfer Regulations.
6. *See* Robert Kneller (1999), 'Ownership Rights to University Inventions in Japan and China', *CASRIP Publication Series: Streamlining Int'l Intellectual Property*, p.162, available at www.law.washington.edu/Casrip/Symposium/Number5/pub5atcl18.pdf.
7. Carl J. Dahlman & Jean-Eric Aubert (2001), *China and the Knowledge Economy – Seizing the 21st Century*, p.124.
8. Shulin Gu (1999), *China's Industrial Technology: Market Reform and Organizational Change*, pp.29–30.
9. In his book, *Capitalism with Chinese Characteristics – Entrepreneurship and the State*, Professor Yasheng Huang makes a similar argument about the impact of China's general reform attitude in the 1980s and its ability to influence China's rural entrepreneurs. Yasheng Huang (2008), *Capitalism with Chinese Characteristics – Entrepreneurship and the State*, pp. 37–38. Professor Hung speculates that rural entrepreneurs during the late 1970s and throughout the 1980s were not necessarily responding to the details of specific political reforms. Instead these entrepreneurs were responding to the perception of economic liberalization that was embodied by Deng Xiaoping, which Professor Hung refers to as the 'Deng Xiaoping effect.' *Id.* at 38.

> 'The importance of Deng is that he was *observably different* from Mao. (And I am not just thinking about their difference in physique.) The key word here is 'observable' – Deng had a set of credential that were not obtuse but commonly known. The case of knowledge is important. The entrepreneurial response originated not from a select group of urban elites but from hundreds of millions of Chinese peasants scattered in far-flung places. They had to believe that the policy change under Deng was permanent rather than cyclical and that Deng's China was objectively different from Mao's China *Id.*

10. Promulgated by Order No. 265 of the State Council of the PRC on May 23, 1999. Revised according to the Decision of the State Council on Amending the Regulation on National Awards for Science and Technology on December 20, 2003.
11. Promulgated by Order No. 1 of the Ministry of Science and Technology on December 24, 1999. Revised in 2004 pursuant to Order No. 9 of the Ministry of Science and Technology on December 27, 2004. Revised again in 2008, pursuant to Order No. 13 of the Ministry of Science and Technology on December 23, 2008.
12. Promulgated by Order No. 29 of the State Council General Office on March 30, 1999 (jointly formulated by the Ministry of Science and Technology, the Ministry of

Education, the Ministry of Personnel, the Ministry of Finance, the People's Bank of China, the State Administration of Taxation, and the State Administration for Industry and Commerce).

13.   The list was motivated by a summary provided by Tang Ming-feng (2006), 'Effect of China's 'Bayh–Dole Act' on University Research and Technology Transfer', 13 *J. Cent. South Univ. Technol.* 10, 10 [hereinafter Tang – *China Bayh–Dole*].

14.   Article 20 of the Law on Science and Technology Progress.

15.   Article 16 of China's Patent Law

16.   Article 20 of the Law on Science and Technology Progress.

17.   *Id.*

18.   Article 1.3 of the Provisions on Promoting the Application of Scientific and Technological Achievements.

19.   Article 1.1 of the Provisions on Promoting the Application of Scientific and Technological Achievements.

20.   Article 1.5 of the Provisions on Promoting the Application of Scientific and Technological Achievements.

21.   *Id.*

22.   Tang – *China Bayh–Dole, supra* note 13, at 11.

23.   *Id.*

24.   Enying Zheng & Hongxing Yang (2007), *Institutionalization of Technology Transfer in Chinese Universities*, paper presented at the annual meeting of the American Sociological Association (Aug. 11), 5, available at www.allacademic.com/meta/p_mla_apa_research_citation/1/8/4/9/1/p184918_index.html.

25.   Wei Hong (2005), *Technology Transfer of Chinese Universities: Forms and Implications*, Paper presented at the annual meeting of the American Sociological Association, (Aug 12), 5, available at www.allacademic.com/meta/p21558_index.html.

26.   Article 9 of the Patent Law.

27.   Hua Guo, *supra* note 2, at 1677.

28.   *Id.*

29.   Nanjing University is ranked 27th in terms of effective patents, so it is not shown in Table 6.3. Nanjing University has 408 effective patents, of which 391 are invention patents. China Education and Research Network (citing to Ministry of Education S&T Development Center), www.edu.cn/shu_ju_pai_hang_1088/20090622/t20090622_385829.shtml.

30.   University of California – Office of Technology Transfer, *University Technology Transfer – Questions and Answers, Part 1: 6. What factors influence university decisions to license patents either exclusively or non-exclusively?*, available at www.ucop.edu/ott/faculty/tech.html.

31.   *Id.*

32.   Wei Hong, *supra* note 25, at 8.

33.   Weiping Wu (2010), 'Managing and Incentivizing Research Commercialization in Chinese Universities', 35 *J. Technol. Transf* 203, 206.

34.   *Id.*

35.   *See id.*

36.   University of California – Office of Technology Transfer, *University Technology Transfer – Questions and Answers, Part 1: 6. What factors influence university decisions to license patents either exclusively or non-exclusively?*, available at http://www.ucop.edu/ott/faculty/tech.html.

37.   *Id.*

38.   *Id.*

39.   *Id.*

40.   Wu, *supra* note 33, at 211.

41.   *See* Wu, *supra* note 33, at 210–212.

42.   The number of non-S&T USUs declined from roughly 3350 in 2000 to roughly 2200 in 2004.

43. Wu, *supra* note 33, at 216.
44. *See* David Barboza (1998), 'Loving a Stock, Not Wisely but too Well', *N.Y. Times*, Sept. 20 at Section 3, p.1.
45. *Id.*
46. *Id.*
47. *Id.*
48. Wu, *supra* note 33, at 216.
49. *See generally id*, at 212.
50. *Id.*, at 208.
51 *See e.g.*, Wei Hong, *supra* note 25, at 8–9.

CHAPTER 7

1. Hu Jintao made these remarks in an October 2007 report to the 17th National Congress of the Communist Party of China. 'Innovation Tops Hu Jintao's Economic Agenda', *Xinhua News* (Oct. 15, 2007), available at http://news.xinhuanet.com/english/2007-10/15/content_6883390.htm.
2. Paul Romer (1990), 'Endogenous Technological Change', 98 *Journal of Political Economy* S71, S74.
3. *See e.g.*, Kenan Patrick Jarboe and Robert D. Atkinson (1998), 'The Case for Technology in the Knowledge Economy – R&D, Economic Growth, and the Role of Government', *Progressive Policy Institute Policy Briefing*, p.5 (June). *See also* Joseph Cortright (2001), 'New Growth Theory, Technology and Learning: A Practitioner's Guide', *U.S. Economic Development Administration Reviews of Economic Development Literature and Practice*, No. 4, p.7, available at www.eda.gov/PDF/1G3LR_7_cortright.pdf.
4. Jarboe and Atkinson, *supra* note 3, at 5–6.
5. *Id.*
6. *See* Chapter 3, Section 4 and Chapter 6, Section 2.
7. *See generally,* Wei Hong (2005), *Technology Transfer of Chinese Universities: Forms and Implications*, paper presented at the annual meeting of the American Sociological Association (Aug 12), p.9, available at www.allacademic.com/meta/p21558_index.html.
8. *See* Chapter 6, Section 2.4.
9. Gregory C. Chow (2007), *China's Economic Transformation*, 2nd ed., p.44.
10. *What is the Five-Year Plan?*, Gov.CN – Chinese Government's Official Web Portal (Apr. 5, 2006), available at http://english.gov.cn/2006-04/05/content_245556.htm.
11. The OECD Frascati manual provides the standard definition for GERD. It defines GERD as the sum of R&D expenditures performed in a country in four economic sectors: (1) business enterprise; (2) government; (3) private non-profit; and (4) higher education. OECD (2002) *Frascati Manual: Proposed Standard Practice for Surveys on Research and Environmental Development*, pp.121–122.
12. Barry Naughton (2007), *The Chinese Economy – Transitions and Growth*, p.352.
13. *Id.*
14. *Id.*
15. *Id.*
16. *Id.*, at 355–356.
17. Albert G.Z. Hu & Gary H. Jefferson (2008), 'Science and Technology in China', in Loren Brandt and Thomas G. Rawski (eds), *China's Great Economic Transformation*, p.295.
18. OECD (2008), *Science Technology and Industry Outlook 2008*, at 80.
19. *Id.*
20. John Whalley & Weimin Zhou (2007), 'Technology Upgrading and China's Growth

Strategy to 2010', *The Centre for International Governance Innovation Working Paper No. 21* available ar www.cigionline.org, at p.14.

21. 2008 STI Outlook, *supra* note 18, at 83.
22. Whalley & Zhou, *supra* note 20, at 15.
23. *China's Unified Enterprise Income Tax Law Brings Greater Clarity, Transparency & Fairness to the Tax System – A Favourable Development for all Market Participants*, Deloitte Press Releases (Mar. 16, 2007), available at www.deloitte.com/dtt/press_release/0,1014,sid%253D7062%2526cid%253D149787,00.html.
24. Whalley & Zhou, *supra* note 20, at 17.
25. *Id.*
26. *See* OECD (2007), *Reviews of Innovation Policy: China – Synthesis Report* 12 [hereinafter *China – Synthesis Report*], available at www.oecd.org/dataoecd/54/20/39177453.pdf, which states: 'China's opening to [FDI] was not motivated by a shortfall of domestic savings.' As evidence for that assertion, an OECD economic survey of China notes that China's 'current account (which measures the difference between domestic saving and investment) has been in surplus for all but one year since 1991.' OECD (2005) *Economic Surveys: China* 35 [hereinafter *China Economic Survey*], available at www.oecd.org/document/21/0,3343,en_2649_34111_35331797_1_1_1_1,00.html.
27. *See China – Synthesis Report*, *supra* note 26, at 12.
28. Whalley & Zhou, *supra* note 20, at 4. *China – Synthesis Report*, *supra* note 26, at 11–12. *China Economic Survey*, *supra* note 26, at 35–36.
29. *China – Synthesis Report*, *supra* note 26, at 11.
30. Whalley & Zhou, *supra* note 20, at 9.
31. *Id.*, at 9–10.
32. Linda Jakobson (2007), 'China Aims High in Science and Technology', in Linda Jakobson (ed.), *Innovation with Chinese Characteristics – High-Tech Research in China*, p.2.
33. *China – Synthesis Report*, *supra* note 26, at 12.
34. Cong Cao, Richard P. Suttmeier & Denis Fred Simon (2006), 'China's 15-Year Science and Technology Plan', *Physics Today*, pp.38, 39 (December), available at www.levin.suny.edu/pdf/Physics%20Today-2006.pdf.
35. *Id.*
36. *Id.*
37. Katherine Connor Linton (2008), *China's R&D Policy for the 21st Century: Government Direction of Innovation*, p.4 (February), available at http://ssrn.com/abstract=1126651.
38. *Id.*
39. Whalley & Zhou, *supra* note 20, at 2
40. *See id.*
41. 'Revised Forecast Advances Date of China Becoming the Preeminent Global Manufacturer', *Global Insight: Our Perspective* (Aug. 12, 2008), available at www.globalinsight.com/Perspective/PerspectiveDetail13718.htm, announcing that its forecasts project that China will overtake the United States as the world's leading manufacturer by 2016–17.
42. *See* Whalley & Zhou, *supra* note 20, at 2.
43. We are purposely over-simplifying the manufacturing sector. There is high-end manufacturing that is not akin to a commodity and is very desirable for a country to attract. However, high-end manufacturing is by no means the majority of manufacturing.
44. Whalley & Zhou, *surpa* note 20, at 2. One reason for the increased manufacturing costs is that Chinese labor is becoming more expensive. *China Economic Survey*, *supra* note 26, at 36.
45. National Development and Reform Commission of the People's Republic of China, *The Outline of the Eleventh Five Year Plan – For National Economic & Social Development of the People's Republic of China*, Chap. 6, available at http://en.ndrc.gov.cn/hot/t20060529_71334.htm.
46. Xu Guanhua, Minister of Science and Technology, at a press conference on March 10,

2006 addressing China's plan to build an innovation-oriented country, 'China Needs 40-Year 7% Annual Growth to Meet Well-Off Target', *China Daily* (Mar. 11, 2006), available at www.chinadaily.com.cn/english/doc/2006-03/11/content_532886.htm.

47. *See generally* Naughton, *supra* note 12, at 366–368, who explains:

> 'Technology development is the unifying thread that links together many aspects of economic policy in the Hu Jintao-Wen Jiabao administration (2002–). A newly emphatic stress on human resources as the foundation of development policy is clearly related, by way of the long-term development of the human capital base, to the promotion of the technology industry. The top priority of foreign trade development has been placed firmly on the development of high-technology trade. Trade promotion policies stress that the key to upgrading exports is the promotion of high-tech exports, particularly those in which China has its own intellectual property, and it is taken for granted that government has a role in promoting such exports. In the sphere of corporate governance . . . high-technology enterprises have been at the cutting edge of changes that give managers stakes in the firms they run. Finally, there is technology policy per se, in which an enormous range of subsidies and financial support packages are available to literally thousands of private, state-owned, and foreign-invested firms'. *Id.*, at 366–367.

We believe that the unifying threads identified by Naughton can be traced back to the early 1980s, and not just back to 2002.

CHAPTER 8

1. Paul A. Gompers & Josh Lerner (2001), *The Money of Invention – How Venture Capital Creates New Wealth*, p.1 [hereinafter *Money of Invention*].
2. *See* Chapter 3, Section 4; and Chapter 6, Section 2.
3. *See* Mark Van Osnabrugge and Robert J. Robinson (2000), *Angel Investing: Matching Start-Up Funds with Start-Up Companies – The Guide for Entrepreneurs, Individual Investors, and Venture Capitalists*, p.22.
4. One technique that has been used to measure the innovation advantage from rapid-growth start-ups is to examine the patents that come out of companies that have received financing from venture capital (VC) funds. *See* Samuel Kortum & Josh Lerner (2000), 'Assessing the Contribution of Venture Capital to Innovation', 31 *Rand J. Econ.* 674, 674–675, 689–691 (finding that VC-backed companies produced more patents than non-VC-backed companies and the patents the VC-backed firms produced were apparently more valuable).
5. *See* Jeffrey E. Sohl (1999), 'The Early-Stage Equity Market in the USA', 1 *Venture Capital* 101, 105. ('Over the last 4 years [from 1996 to 1999], these high growth start-ups added 6 million jobs to an economy that added 7.7 million jobs in total. *Id*; *see also* David Birch et al. (1994), *Who's Creating Jobs?*, 6–7:

> 'Most of the new jobs attributable to small firms are thus created by a relatively few small firms that start small and grow fast. Said another way, most small firms grow slowly. It is not the local drug store or beauty shop or restaurant that is the main engine of job growth – it is the Gazelle' [Birch's nickname for 'mostly smaller firms that start with the intent to grow, and pull it off'].

> *Id.* Gazelles (which accounted for no more than 3% of firms) added 4.4 million jobs to the U.S. economy between 1989 and 1993, a period when the economy hardly grew. *Id.* at 6.

6. *See* Daniel Sandler (2004), *Venture Capital and Tax Incentives: A Comparative Study of Canada and the United States*, p.2.

7.   Jonsson Yinya Li (2005), *Investing in China: The Emerging Venture Capital Industry*, p.2.
8.   *Money of Invention, supra* note 1, at 11.
9.   A securable asset refers to those assets of a company that can be pledged as collateral to support a loan.
10. William Murphy (2002), 'Proposal for a Centralized and Integrated Registry for Security Interests in Intellectual Property', 41 *IDEA* 297, 297 (proposing the creation of a centralized or integrated registry to ease the perfecting of collateral interests in intellectual property rights); *see also Money of Invention, supra* note 1, at 6.
11. *See* Paul Gompers and Josh Lerner (2006), *The Venture Capital Cycle*, 2nd Ed., pp.157–158 [hereinafter *Venture Capital Cycle*].
12. *Id.*, at 158.
13. Murphy, *supra* note 10, at 306.
14. Georges Doriot, who is viewed by many as the founder of modern venture capital, is credited as saying something along the following lines regarding the importance of management execution: '[a]lways consider investing in a grade A man with a grade B idea. Never invest in a grade B man with a Grade A idea.' William D. Bygrave, (2004) 'The Entrepreneurial Process', in William D. Bygrave & Andrew Zacharakis (eds) *The Portable MBA In Entrepreneurship*, 3rd Ed., 12.
15. Adam Smith described the problem as follows:

     'The directors of [joint-stock] companies, however, being the managers rather of other people's money than of their own, it cannot well be expected, that they should watch over it with the same anxious vigilance with which the partners in a private copartnery frequently watch over their own. Like the stewards of a rich man, they are apt to consider attention to small matters as not for their master's honour, and very easily give themselves a dispensation from having it. Negligence and profusion, therefore, must always prevail, more or less, in the management of the affairs of such a company.'

     Adam Smith, 'Of the Revenue of the Sovereign or Commonwealth', in *An Inquiry into the Nature and Causes of the Wealth of Nations*, 264–265.
16. Money of Invention, *supra* note 1, at 44–59.
17. Martin Kenney, Martin Haemmig & W. Richard Goe, 'The Globalization of the Venture Capital Industry', *Innovation in Global Industries: U.S. Firms Competing in a New World (Collected Studies)* (eds. David C. Mowery & Jeffrey T. Macher), 313. Financiers of entrepreneurial companies during the United States' post-World War II economic expansion met with considerable success and eventually began to involve outsiders in the investment process. *Money of Invention, supra* note 1, at 88. In 1946, the first 'truly modern' VC firm, American Research and Development ('ARD'), was created by 'MIT president Karl Compton, Harvard Business School professor Georges F. Doriot, and several local business leaders . . . to make high-risk investments in emerging companies that based their innovations on technology developed from the war effort.' *Id.* ARD and other early venture capitalists focused on young technology-based start-ups (including most notably in the minicomputer and the semiconductor industries), and that focus on young technology companies continues today. *Id.*
18. Kenney, Haemmig & Goe, *supra* note 17, at 315. In Japan and Korea, for example, venture capital investing has 'traditionally been in the form of loans.' *Id.*
19. *Id.*
20. The term 'angel' originated in the early 1900s to refer to wealthy backers of Broadway shows who made risky investments to support these productions. Gerald A. Benjamin & Joel Margulis (1996), *Finding Your Wings: How to Locate Private Investors to Fund Your Venture*, p.5.
21. For a detailed discussion of the U.S. angel market, see John L. Orcutt (2005),

'Improving the Efficiency of the Angel Finance Market: A Proposal to Expand the Intermediary Role of Finders in the Private Capital Raising Setting', 37 Ariz. St. L.J. 861, 876–879 [hereinafter Orcutt–*Angel Finance Market*].

22. *Money of Invention, supra* note 1, at 23.
23. *Id.*
24. For technology companies, this entails understanding the strength and viability of any new and proprietary technology being developed by the company.
25. It is common that a rapid-growth start-up's primary assets are intellectual property. As a result, it is critical for investors to understand the strength of those rights. For example, if the start-up's product is based on technology licensed from another party, what are the parameters/restrictions of that license? If the value of the start-up depends on its ability to patent its proprietary technology, what is the anticipated strength of those patents?
26. Because the value of the company will generally depend on the expected future profitability of the company, projections (including how they were developed and their reasonableness) are a fundamental part of the due diligence process.
27. *See Venture Capital Cycle, supra* note 11, at 157–163. *See also* Bernard S. Black and Ronald J. Gilson (1998), 'Venture Capital and the Structure of Capital Markets: Banks versus Stock Markets', 47 *Journal of Financial Economics* 243, 252–255.
28. *Id.*, at 252.
29. *Id.*, at 252–253.
30. In the United States, for example, state corporate statutes typically contain a code section that provides that corporations are to 'be managed by or under the control of a board of directors.' *E.g.*. Del. Code Ann. Tit. 8, § 141(a); *see also* Model Bus. Corp. Act § 8.01(b) ('All corporate powers shall be exercised by or under the authority of, and the business and affairs of the corporation managed by or under the direction of, its board of directors . . . '); *see* Franklin A. Gevurtz (2000), *Corporation Law, Hornbook Series*, 190 for the proposition that most states have a similar state corporate code provision. As a result, the board is the ultimate decision making authority for a corporation.

Chinese law is comparable to U.S. state law on this front. Articles 47 and 109 of China's Company Law similarly provide for the preeminence of the board of directors in China's two primary 'corporate-type' entities – the limited liability company and the joint stock limited company.

Article 47 of China's Company Law states:

'The board of directors [for a limited liability company] shall be responsible for the shareholders' meeting and exercise the following authorities:

1.   Convening shareholders' meetings and reporting the status on work thereto;
2.   Carrying out the resolutions made at the shareholders' meetings;
3.   Determining the operation plans and investment plans;
4.   Working out the company's annual financial budget plans and final account plans;
5.   Working out the company's profit distribution plans and loss recovery plans;
6.   Working out the company's plans on the increase or decrease of registered capital, as well as on the issuance of corporate bonds;
7.   Working out the company's plans on merger, split-up, change of the company form, dissolution, and etc.;
8.   Making decisions on the establishment of the company's internal management departments;
9.   Making decisions on hiring or dismissing the company's manager and his remuneration, and, according to the nomination of the manager, deciding on the hiring or dismissing of vice manager(s) and the person in charge of finance as well as their remuneration;

10.   Working out the company's basic management system; and
11.   Other functions as prescribed in the articles of association.'

Article 109 of China's Company Law states:

'A joint stock limited company shall set up a board of directors . . . The provisions in Article 47 of this Law on the functions of the board of directors of a limited liability company shall apply to that of the board of directors of a joint stock limited company.'

31.   Black and Gilson, *supra* note 27, at 254.
32.   *Id.*
33.   *Id.*
34.   For an overview of financial intermediaries, *see* Stephen J. Choi (2004), 'A Framework for the Regulation of Securities Market' Intermediaries, 1 *Berkeley Bus L.J.* 45; Orcutt–*Angel Finance Market*, *supra* note 21, at 884–888. For a specific discussion of research analysts and their role as financial intermediaries, *see* John L. Orcutt (2003), 'Investor Skepticism v. Investor Confidence: Why the New Research Analyst Reforms Will Harm Investors', 91 *Denv. U. L. Rev.* 1 (2003).
35.   *See infra* Table 8.4.
36.   *See* Steven White, Jian Gao and Wei Zhang (2004), 'Financial Systems, Investment in Innovation, and Venture Capital: The Case of China', in Anthony Bartzokas and Sunil Mani (eds), *Financial Systems, Corporate Investment in Innovation, and Venture Capital*, 172–173.
37.   *Id.* Takeshi Jingu & Tetsuya Kamiyama, China's Private Equity Market, 11 *Nomura Capital Market Review* 24, 25; Wei Xiao (2002), 'The New Economy and Venture Capital in China', *Perspectives* Vol. 3 No. 6, available at www.oycf.org/oycfold/http-docs/Perspectives2/18_093002/Economy_Venture_China.htm.
38.   Kuntara Puktuanthong & Thomas Walker, 'Venture Capital in China: A Culture Shock for Western Investors', 45 *Management Decision* 708, 711. Jingu & Kamiyama, *supra* note 37, at 25. Wei Xiao, *supra* note 37.
39.   David Ahlstrom, Garry D. Bruton & Kuang S. Yeh, 'Venture Capital in China: Past, Present, and Future', 24 *Asia Pacific J. Manage* 247, 250 (2007).
40.   *Id.*
41.   *Id.*
42.   Zero2IPO (subscription needed) (statistics on file with authors).
43.   *Id.*
44.   Pukthuanthong & Walker, *supra* note 38, at 711.
45.   *Id.*
46.   *See* White, Gao & Zhang, *supra* note 36, at 173.
47.   *Id.*, at 181.
48.   Li, *supra* note 7, at 33.
49.   White, Gao & Zhang, *supra* note 36, at 173.
50.   *Id.*, at 181.
51.   Jack C. Fensterstock and Aimin Li, *Status of Venture CApital in China 2*, 4–5 (2000), available at www.mcbc.net/Articles/China%20VC%20%20Status.pdf.
52.   White, Gao & Zhang, *supra* note 36, at 181.
53.   Fensterstock & Li, *supra* note 51, at 4.
54.   *See Money of Invention*, *supra* note, at 15.
55.   *See* Claudio Michelacci & Javier Suarez (2004), 'Business Creation and the Stock Market', 71 *Review of Economic Studies* 459, 460 (2004).
56.   James H. Lokey, Jr. and Donald E. Rocap (2008), 'Selected Tax Issues in Structuring Private Equity Funds', 841 *PLI/Tax* 741, 754.
57.   David H. Hsu & Martin Kenney, *Organizing Venture Capital: The Rise and Demise of American Research & Development Corporation, 1946–1973* (Dec. 1, 2004) 29, available at http://ssrn.com/abstract=628661.
58.   White, Gao & Zhang, *supra* note 36, at 171.

59. *Id.*, at 172.
60. Shulin Gu (1999), *China's Industrial Technology: Market Reform and Organizational Change*, 83.
61. White, Gao & Zhang, *supra* note 36, at 171-172.
62. Gu, *supra* note 60, at 352.
63. *Id.*
64. White, Gao & Zhang, *supra* note 36, at 171–172.
65. *Id.*, at 172.
66. *Id.*
67. *Id.*, at 171–172.
68. Sandler, *supra* note 6, at 14.
69. *See generally* White, Gao & Zheng, *supra* note 36, at 173.
70. *Id.*
71. *See* Feng Zeng (2004), *Venture Capital Investments in China* 30, Published Ph.D. dissertation, The Pardee RAND Graduate School, available at http://www.rand.org/pubs/rgs_dissertations/RGSD180/.
72. Li, *supra* note 7, at 39. White, Gao and Zhang, *supra* note 36, at 181.
73. *Id.*
74. *Id.*
75. *Id.*, at 181 and 190.
76. *Id.*, at 181.
77. *Id.*
78. *Id.*
79. *See generally Money of Invention, supra* note 1, at 15.
80. *See* White, Gao and Zhang, *supra* note 36, at 181.
81. *See e.g.*, the Regulations for the Implementation of the Law on Chinese–Foreign Equity Joint Ventures (1983) (English version available at http://fdi.gov.cn/pub/FDI_EN/Laws/law_en_info.jsp?docid=51012); and Rules for the Implementation of the Law on Foreign-Capital Enterprises (1990) (English version available at www.law-lib.com/lawhtm/1990/52600.htm).
82. White, Gao and Zhang, *supra* note 36, at 173. Prior to 1992, most FVC investments were conducted through China Direct Investment Funds. *Id. See also* Ahlstrom, Bruton & Yeh, *supra* note 39, at 250.
83. White, Gao and Zhang, *supra* note 36, at 191.
84. *Id.*, (citing to Wei Zhang & Yanfu Jiang (2002), *The Relationship Between Venture Capitalists' Experience and their Involvement in the VC-Backed Companies*, paper presented at the Global Finance Corporation, Guanghua School of Management, Beijing University (May 27–29)).
85. *See id.*, at 181.
86. Article 4 of the FVC Regulations.
87. *Id.*
88. *Id.*
89. Jingu & Kamiyama, *supra* note 37, at 29.
90. Article 3 of the 2003 FVC Regulations.
91. Article 31 of the 2003 FVC Regulations.
92. Articles 3 and 31 of the 2003 FVC Regulations.
93. Article 31(2) of the 2003 FVC Regulations.
94. Article 32 of the 2003 FVC Regulations.
95. *Id.*
96. Article 8 of the 2003 FVC Regulations. Article 8 calls for approval by the Ministry of Foreign Trade and Economic Cooperation (MOFTEC), but MOFTEC has since been subsumed by the Ministry of Commerce.
97. Article 6 of the 2003 FVC Regulations.
98. *Id.*
99. Article 13 of the 2003 FVC Regulations.

100. *Id.*
101. Article 7 of the 2003 FVC Regulations
102. *Id.*
103. Article 17 of the 2003 FVC Regulations.
104. Article 21 of the 2003 FVC Regulations.
105. *Id.*
106. Article 37 of the 2003 FVC Regulations.
107. White, Gao and Zhang, *supra* note 36, at 183.
108. *See id.*, at 195.
109. *See e.g., id.* at 193–194; Lu Haitian, Tan Yi & Chen Gongmeng (2007), 'Venture Capital and the Law in China', 37 *Hong Kong L. Jour.* 229, 242–243 (2007).
110. Justin Tan, Wei Zhang & Jun Xia (2008), 'Managing Risk in a Transitional Environment: An Exploratory Study of Control and Incentive Mechanisms of Venture Capital Firms in China', 46 *Journal of Small Business Management* 263.
111. *Id.*, at 274–275.
112. *Id.*, at 275.
113. Rather than provide a start-up with the capital needed to fund its entire business plan, a venture capital firm is likely to 'stage' its investments through various rounds, with each round of financing intended to finance the company to a particular milestone or milestones. Staging the investments provides the venture capital firm with an 'option to abandon' the investment, which can be a powerful tool to reduce both information and agency problems. To begin with, staging helps to reduce the uncertainty problem. For example, if the venture capital firm is concerned about the viability of the company's technology or its ability to achieve certain milestones, staging the investment allows the company to risk only a portion of the investment on the company up front. The initial investment may be of an amount to allow the venture capital firm to become more comfortable with the technology or to allow the company to reach some of the milestones. Staging the investments also reduces information asymmetries by providing venture capital firms with improved access to the company's most confidential information. Prior to becoming an investor, the company's management has substantially greater access to crucial information about the investment-worthiness of the company, such as the reliability of the company's financial projections and the true capabilities of management. Making a partial initial investment allows the venture capital firm to obtain similar access as management to much of this information. Ronald J. Gilson (2003), 'Engineering a Venture Capital Market: Lessons from the American Experience', 55 *Stan. L. Rev.* 1067, 1073, 1078–1081.
    Staging also helps to reduce agency concerns for the venture capital firm. Staging places the decision of whether to continue to fund the company in the hands of the venture capital firm, which shifts a substantial amount of discretion regarding the direction of the company from its managers to its owners (i.e., the venture capital firms). Requiring management to remain beholden to the venture capital firm for additional funding greatly reduces the ability of the company's managers to act strategically. *Id.*
114. Tan, Zhang & Xia, *supra* note 110, at 281.
115. Shaun Rein (2008), 'Progress or Pipe Dreams? Private Equity/Venture Capital in China'. *J.P. Morgan's Hands-On China Series* (May 2008), available at http://seekingalpha.com/article/76036-progress-or-pipe-dreams-private-equity-venture-capital-in-china.
116. For a thoughtful discussion on that consideration, *see e.g., id.*
117. *Id.*
118. The Employee Retirement Income Security Act of 1974, Pub. L. No. 93-406, 88 Stat. 829 (codified in scattered sections of 5, 18, 26, 29, 31, 42 U.S.C.).
119. 29 U.S.C.A. § 1104 (West Supp. 2005).
120. 29 C.F.R. §§ 2550.404.
121. Article 43 of the Law on Commercial Banks (effective Dec. 27, 2003).
122. Article 105 of the Insurance Law (effective Oct. 28, 2002).

123. Circular on the Regulation of Securities Companies Engaged in Venture Capital Investments (2001).
124. Article 25 of the Interim Measures on the Management of National Social Securities Funds (effective Dec. 13, 2001).
125. 'A Legal Perspective on China's Venture Capital Rush', *China Journal* (2006), available at www.lexsina.ch/uploads/media/Chinajournal_mar.2006_Venture_Capital.pdf [hereinafter *China's VC Rush*].
126. *See* Ada Pearson (2009), *RMB-denominated Fund Investment Evolves into Mainstream on China's VC/PE Market*, Zero2Ipo Research Center (Mar. 28).
127. Li, *supra* note 7, at 20.
128. *See*, John Downes and Jordan Elliot Goodman (1985), *Dictionary of Finance and Investment Terms*, 3rd Ed, p.40.
129. *China's VC Rush, supra* note 125.
130. Li, *supra* note 7, at 22.
131. *China's VC Rush, supra* note 125.
132. Neal Stender, Yan Zeng, Xiaowei Yin & Ying Zhu (2005), 'It's SAFE Again for China Venture Capital Round-trips & Red-chip Listings', *China Law & Practice* 15, 15 (Nov.).
133. The liquidation of a company refers generally to the winding up, or ending, of the company in its current form. The liquidation can occur because the company is sold to another entity or is shut down.
134. *Money of Invention, supra* note 1, at 55.
135. *See id.*
136. *Id.* at 55–58.
137. Participation rights refer to additional liquidation rights that may be granted to preferred stockholders. Preferred stock is referred to as having 'participation rights' when the preferred stockholders, after receiving their full liquidation preference, are then entitled to 'participate' with the common stockholders in any additional amounts distributed to stockholders.
138. A mandatory redemption right is a right for the investor to require the company to repurchase its stock at a certain price and during a certain time period.
139. Common special voting rights include class voting rights on business combinations or new financings.
140. Anti-dilution rights increase the amount of stock the preferred stockholders receive in the event that a future financing is done at a lower valuation.
141. Black & Gilson, *supra* note 27, at 258–259.
142. Haitian, Yi & Gongmeng, *supra* note 109, at 261.
143. Venture Capital Cycle, *supra* note 11, at 345.
144. *See* Armin Schwienbacher (2008), 'Innovation and Venture Capital Exits', 118 *Economic Journal* 1888, 1888; D. Gordon Smith (2005), 'The Exit Structure of Venture Capital', 53 *UCLA L. Rev.* 315, 316.
145. Matthias Eckermann (2005), *Venture Capitalists' Exit Strategies under Information Asymmetry: Evidence from the US Venture Capital Market*, p.3.
146. *Id.*
147. Schwienbacher, *supra* note 144, at 1888.
148. Sandler, *supra* note 6, at 10.
149. Simply put, there are more parties that would like to receive investment capital than there is available investment capital. In order to be motivated to part with their investment capital, these sources of capital will charge the potential users of capital a price (the cost of capital) to compensate for its use.
150. Shannon P. Pratt (2001), *Business Valuation: Discounts and Premiums*, p. 79.
151. *Id.*
152. Eckermann, *supra* note 145, at 47.
153. 'The need to exit is the core of the venture model. You might have a wonderful business idea, that will make its customers, employees, and managers all happy, but if it

is unlikely to exit, and with a sufficient return on the capital put in, we don't want it. It could in fact generate wonderful cash flow forever, but if we can't monetize that in a liquid form by the end of the fund, it won't make it through the partners' meeting. It is inherent in the nature of the beast.' Tim Oren (a long-time Silicon Valley venture capitalist), 'No Exit: When Venture Capital Isn't Right', *Tim Oren's Due Diligence: Letters from an inhabited dataspace* (May 13, 2003), available at http://due-diligence. typepad.com/blog/2003/05/no_exit_when_ve.html.

154.   *See* Mechelacci & Suarez, *supra* note 55, at 71; and Paul Gompers, Anna Kovner, Josh Lerner & David Scharfstein, 'Skill vs. Luck in Entrepreneurship and Venture Capital: Evidence from Serial Entrepreneurs', *NBER Working Papers 12592* (2006).

155.   Self-liquidating funds are the most common structure in the United States, and presumably are the primary structure for FVCs operating in China.

156.   Lokey & Rocap, *supra* note 56, at 754.

157.   Hsu & Kenney, *supra* note 57, at 29.

158.   The following table provides data for the first half of 2009.

	Venture Capital/PE-Backed Acquisition Exits			Venture Capital/PE-Backed IPO Exits		
	Total	Total with Disclosed Values	Total Value of Disclosed Acquisition Exits (US$ millions)	Total	Total IPO Offer Amount (US$ millions)	Average Size of IPO (US$ millions)
First-half of 2009	17	13	531.79	7	1918.7	274.1

Source: Zero2IPO database.

159.   Black and Gilson, *supra* note 27, at 245.

160.   *Id.*

161.   The belief that IPO exits are more profitable than acquisition exits may be biased by the fact that venture capital firms tend to select their strongest portfolio companies for IPO exits. Eckermann, *supra* note 145, at 65.

162.   *See e.g.*, Douglas Cumming & Jeffrey MacIntosh (2004), 'Boom, Bust, and Litigation in Venture Capital Finance', 40 Willamette L. Rev. 867, 877–878.

163.   *See Money of Invention*, *supra* note 1, at 221.

164.   *Id.*, at 219.

165.   Black and Gilson, *supra* note 27.

166.   *Id.*, at 258–261.

167.   *Id.*, at 258.

168.   *Id.*, at 258–259.

169.   *Id.*, at 259.

170.   *Id.*

171.   ChiNext – Driving Innovation and Growth, Shenzen Stock Exchange, available at www.szse.cn/main/en/ [hereinafter ChiNext Website].

172.   *Id.* – Market News – 28 Companies to be listed on ChiNext Board (Oct. 30, 2009).

173.   *Id.* – Statistics.

174.   Ewing Marion Kauffman Foundation (2007), *On the Road to an Entrepreneurial Economy: A Research and Policy Guide* 28 (July 2007), available at www.kauffman. org/uploadedFiles/entrepreneurial_roadmap_2.pdf.

175.   John L. Orcutt (2009), 'The Case Against Exempting Smaller Reporting Companies from Sarbanes–Oxley Section 404: Why Market-Based Solutions are Likely to Harm Ordinary Investors', 14 *Fordham J. Corp. & Fin. L.* 325, 357.

176. *Id.*
177. *Id.*, at 405–406.

CHAPTER 9

1.   Weiping Wu (2010), 'Managing and Incentivizing Research Commercialization in Chinese Universities', 35 *J. Technol. Transf.* 203, 204.
2.   *See* Chapter 3, Section 4 and Chapter 6, Section 2.
3.   *See* Chapter 6.
4.   *See* Linda Jakobson (2007), *China Aims High in Science and Technology*, 23 and Jun Yu (2007), 'Biotechnology Research in China – A Personal Perspective', 161 in Linda Jakobson (ed.), *Innovation with Chinese Characteristics – High-Tech Research in China* .
5.   OECD (2007), *OECD Reviews of Innovation Policy: China*, p.192.
6.   *See* Wu, *supra* note 1, 207.
7.   Enying Zheng & Hongxing Yang (2007), *Institutionalization of Technology Transfer in Chinese Universities*, paper presented at the annual meeting of the American Sociological Association (Aug. 11), p.4, available at www.allacademic.com/meta/p_mla_apa_research_citation/1/8/4/9/1/p184918_index.html.
8.   *Id.*, at 5–7.
9.   *Id.*, at 5.
10. Robert Kneller (1999), 'Ownership Rights to University Inventions in Japan and China', *CASRIP Publication Series: Streamlining Int'l Intellectual Property*, p.163, available at www.law.washington.edu/Casrip/Symposium/Number5/pub5atcl18.pdf.
11. *Id.*
12. In addition to individual universities developing internal TTO capacity, the Ministry of Education and the former State Economic and Trade Commission authorized six universities in 2001 to establish National Technology Transfer Centers (NTTCs). Ming-feng Tang (2006), *A Comparative Study on the Role of National Technology Transfer Centers in Different Chinese Universities,* paper presented at Globelics India 2006 Conference on Innovation Systems for Competitiveness and Shared Prosperity in Developing Countries (Oct), 1, available at www.globelicsindia2006.org/sessions.php. The six universities were Tsinghua University, Shanghai Jiaotong University, China East Polytechnic University, Huazhong S&T University, Xi'an Jiaotang University and Sichuan University. *Id.* The general role of an NTTC is to:

- Establish joint university/industry R&D efforts;
- Promote university incubation efforts;
- Improve the commercialization of university inventions;
- Serve as an interface between Chinese and international technology firms in order to strengthen their technology cooperation efforts; and
- Provide a variety of services to technology firms.

Id., at 2.
The role of NTTCs substantially overlaps with the basic role of individual, university TTOs. Technically, the biggest difference is the role in international cooperation that was given to NTTCs. *Id.* Because NTTCs are an experimental project, the six universities that were given NTTCs also maintained their existing TTOs – with the two offices co-existing. *Id.* While NTTCs have received some attention from academic researchers, it does not appear that there is anything truly special about NTTCs that differentiates them from the more typical TTOs. As a result, our discussion of university IP management uses the term TTO, which will include both TTOs and the six NTTCs.
13. *Id.*, at 1.
14. *See* Zheng & Yang, *supra* note 7, at 7.
15. *Id.*

16. *Id.*
17. *See e.g.*, Hua Guo (2007), 'IP Management at Chinese Universities', in Anatole Kratiger et al. (eds), *Intellectual Property Management in Health and Agricultural Innovation – a Handbook of Best Practices*, pp.1677–1679.
18. John L. Orcutt (2005), 'Improving the Efficiency of the Angel Finance Market: A Proposal to Expand the Intermediary Role of Finders in the Private Capital Raising Setting', 37 *Ariz. St. L. Rev.* 861, 884.
19. *See* the AUTM website at www.autm.net/home.htm.
20. Based on conversations we had with the TTO director for an elite technology university.
21. Hu Angang (2003), 'How Will China Build a Well-Off Society for All of Its Citizens', *The China Business Review* (Mar.–Apr.), pp.50–55.
22. *See* Ross Gittell & John Orcutt (2010), *New Hampshire in the Innovation Economy: A Plan to Increase Innovation and Technology-Based Economic Development in New Hampshire*, p.17, available at www.epscor.unh.edu/.
23. *Dartmouth Entrepreneurial Network – Welcome to the DEN*, available at www.den. dartmouth.edu/.
24. Gittell & Orcutt, *supra* note 22.
25. *Dartmouth Entrepreneurial Network – Overview*, available at www.den.dartmouth.edu/ about/overview.html.
26. The Association of University Technology Managers (AUTM), *U.S. Licensing Activity Survey – Survey Summary* – FY 2007.
27. Orcutt, *supra* note 18, at 871.
28. Jeffrey E. Sohl (2003), The U.S. Angel and Venture Capital Market: Recent Trends and Developments, *J. Private Equity* 13 (Spring 2003); *see also* Mark Van Osnabrugge & Robert J. Robinson(2002) *Angel Investing: Matching Start-Up Funds with Start-Up Companies – The Guide for Entrepreneurs, Individual Investors and Venture Capitalists*, 47–52 (2000); U.S. GEN. Accounting Office (2000), *Small Business: Efforts to Facilitate Equity Capital Formation* 10 [hereinafter Gao Report].
29. Orcutt, *supra* note 18, at 871–872.
30. Jeffrey E. Sohl (1999), 'The Early-Stage Equity Market in the USA', 1 *Venture Capital* 101, 108.
31. John Freear, Jeffrey E. Sohl & William E. Wetzel, Jr. (1996), *U.S. Small Bus. Admin., Creating New Capital Markets from Emerging Ventures*, p.8.
32. Jeffrey A. Timmons & Harry J. Sapienza (1992), 'Venture Capital: The Decade Ahead', in Donald L. Sexton and John D. Kasarda (eds), *The State of the Art of Entrepreneurship*, pp.402, 421.
33. Colin Mason & Richard Harrison (1996), 'Stimulating 'Business Angels'', in 4 *Venture Capital and Innovation* 54, 61 (Organisation for Economic Co-operation and Development, Working Paper No. 98).
34. John Freear, Jeffrey E. Sohl & William E. Wetzel, Jr (1994), 'The Private Investor Market for Venture Capital', 1 *The Financier: A CMT* 7, 9. The angel market is often referred to as the 'invisible' venture capital market because of angel investors' preference for anonymity. *Id.*

CONCLUSION

1. Henry De Soto (2000), *The Mystery of Capital – Why Capitalism Triumphs in the West and Fails Everywhere Else*, p.228.
2. Justin Yifu Lin (2005), 'Viability, Economic Transition and Reflection on NeoClassical Economics', 58 *KYKLOS* 239, 240.
3. Douglas W. Arner (2007), *Financial Stability, Economic Growth, and the Role of Law*, pp.16–17.
4. *See e.g.*, Linda Jakobson (2007), 'China Aims High in Science and Technology', in Linda Jakobson (ed.) *Innovation with Chinese Characteristics – High-Tech Research in China*. p.27

Index